HOBBY-VAC 12X18

Construction Plans

Build it Yourself and Save !

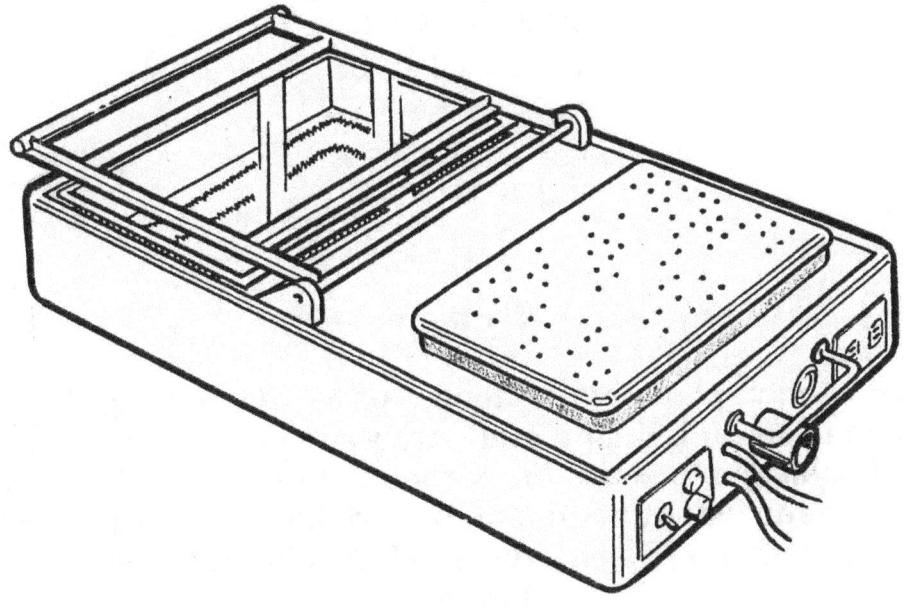

Construction Plans for a 12 x 18 inch. Vacuum Forming Machine with 110 volt Oven

DIY Vacuum Forming

Build plans for the Hobby-Vac plastic molding machine

By Douglas E.Walsh

Second Edition, Copyright 2024
All Rights Reserved, printed in the USA
ISBN# 979-8-9901179-0-7

Workshop Publishing
2909 Crozier Rd.
Lupton, MI 48635
www.build-stuff.com
hobbyvac@yahoo.com
(248)391-2974

Warning!

Contents

11 - Alternate Heating Elements

Finding the resistance wire and insulation boards - Using a two layer design with common cement tile backer board.

12 - Electrical wiring

Choosing and mounting the components - Using a safety timer - Managing loads on the circuit - Safety cautions - Proper grounding - Testing it out

13 - Using your machine

Assembly checklist - Operating sequence - How much to heat the plastic - Heat management.

14 - Plastics and Molds

Properties of common Plastics - Choosing and finding a plastic - Dealing with moisture absorption - Mold making types, materials and sources. Forming tips.

15 - Making your molds

Rules and Tips - Webbing - Undercuts - Practical limits for mold shape and size.

16 - 3D Printed Molds

Advice on choosing the best type of printer and materials for mold making. How to optimize the settings. Dealing with surface finishes

Chapter 1 - What is Vacuum Forming

1 - Vacuum forming or Thermoforming is a process where you start with a flat sheet of plastic held in a clamp frame. Place the plastic under or over a heater until it's soft and pliable.

Your Mold is placed onto the forming surface.

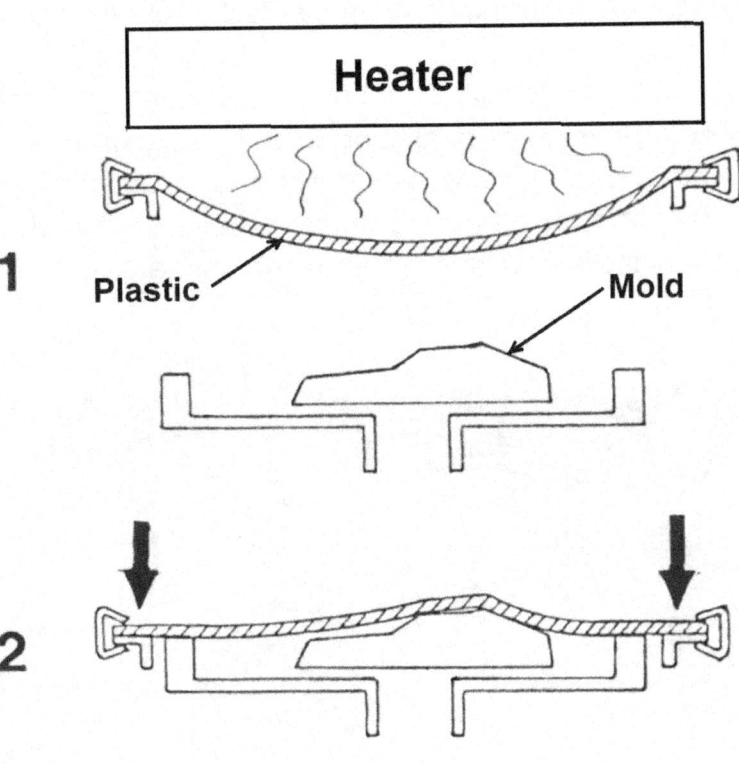

2 - When ready, move the plastic and stretch it over your mold until it forms a seal

3 - Apply vacuum which pulls the soft sheet into corners and depressions leaving you with a faithful copy of your mold.

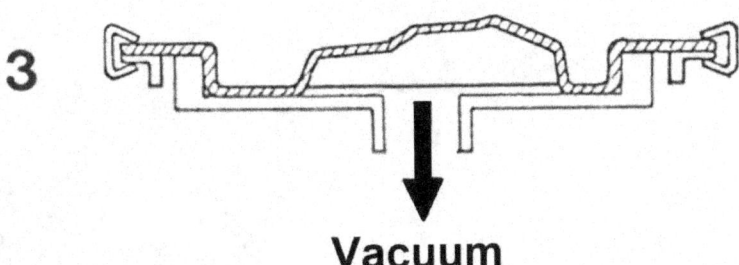

This process is fast and easy... a couple minutes for heating and a couple more for cooling after it's been formed. The plastic can be removed and it will hold its new shape. Trim away the excess and you have a nice part. Reload the frame with another sheet and do it again. Use one mold to make many parts.

What can I make with this Machine?

Let's start right off with some examples. The object on the right is the mold. You can use many materials but this one happens to be a clear plastic car shaped container that came filled with candy. I ate all the candy then I filled it instead with expanding foam to support it. A strong vacuum can distort or crush a weak mold.

The thin plastic sheet on the left was formed directly over it. The process took under three minutes beginning to end. You can use many found items like this for molds but if they are not sturdy you may need to reinforce them. With thin plastic sheets, not much heat is transferred to the mold because it cools so fast. You can use wood, plaster, plastic and many other common materials for molds.

Clear parts

Clear parts,.. no problem. You can make different size platens to use smaller sheets.. You can also fit multiple molds on the full 12x18 inch sheet if you have the room. These are model airplane canopies made from clear PET-G plastic

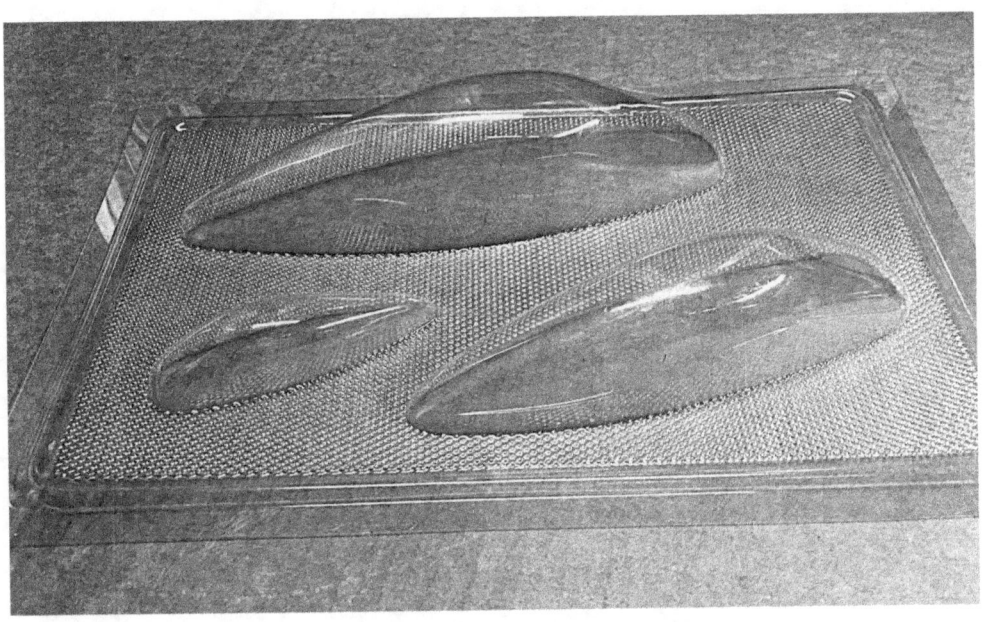

Thick Plastics

The Hobby-Vac can use a real vacuum pump to make it five times more powerful than a shop-vac. You can form up to ¼ inch thick sheets in some types of plastic and still get sharp definition. This is important whether you need thick parts or not. Its an indicator of forming power and more is better. If using thinner plastics it gives you much sharper definition and makes the process less sensitive to proper heating or judging when it's ready. You

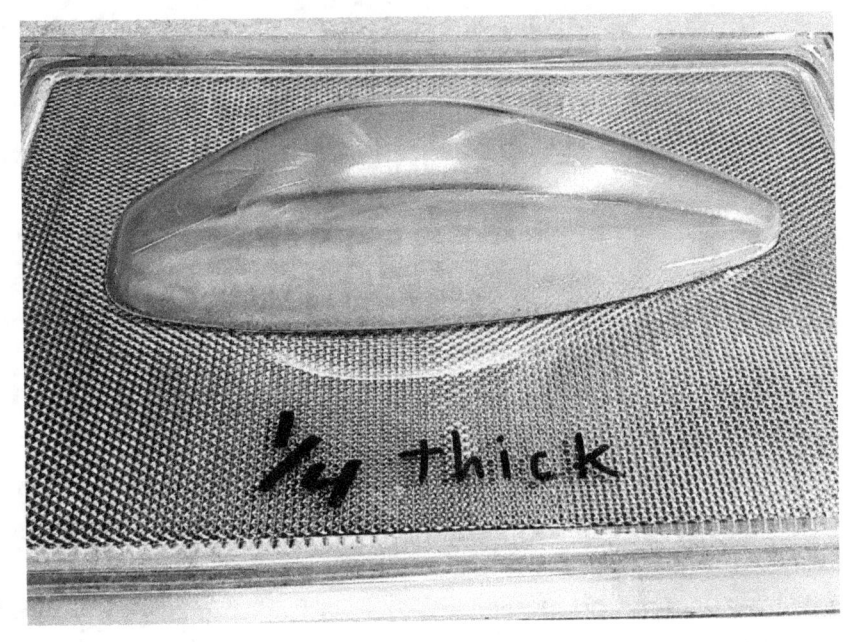

have the power to make it work with less than perfect heat or timing. Too often people have to overheat or over sag the plastic to compensate for a weak vacuum system.

Extreme Detail

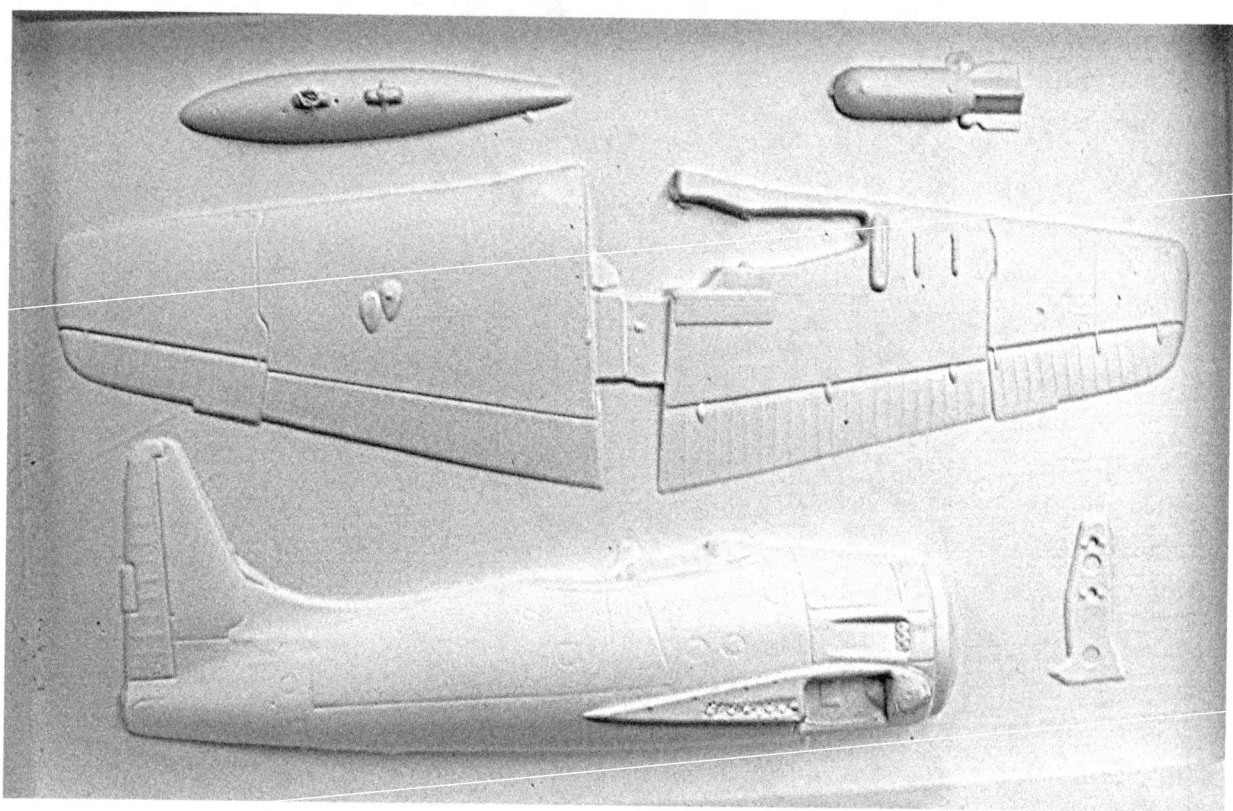

Above is a close up of a part that was formed into a cavity or female mold. You can see that it displays better surface definition with sharper inside corners. This is because you are looking at the surface that was touching the mold. More commonly parts are formed over an object using a male mold. This means the best surface detail is lost on the inside. The mold above was created by pouring casting resin over some model airplane parts creating a block with cavities. Of course small holes are needed in the deepest areas.for air evacuation.

Vacuum Forming is a simple process that can be done easily but the real potential shows up when you get all the parameters right. Part quality and definition depend not only on mold quality but also forming power. These Hobby-Vac plans have the engineering baked in. The internet is full of home built machines that look similar but don't measure up. Follow the plans and you will get it all right. Optimized Infra-red heating and a strong vacuum system are the basics. Follow that up with the rapid response of an airtight low internal volume platen as shown in the plans.

Large Parts - Deep Draws

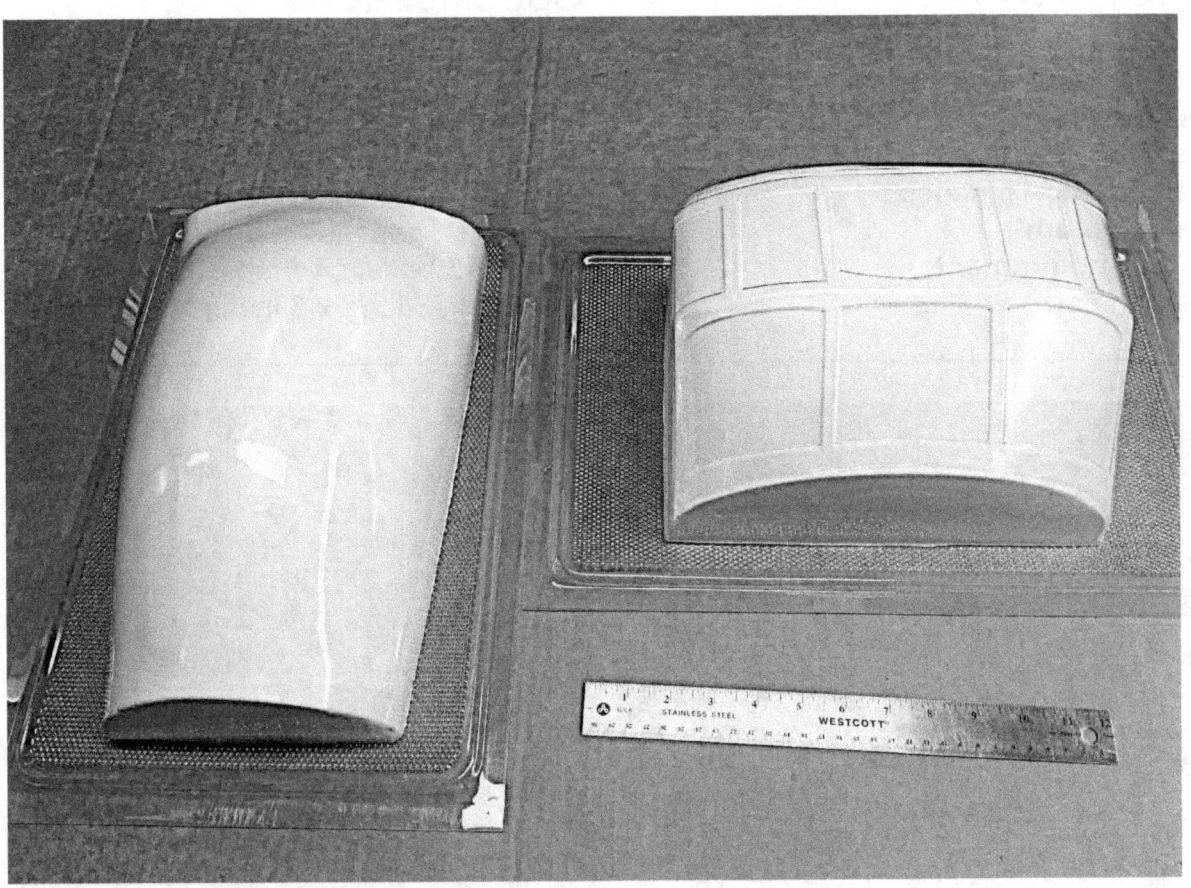

Shown above are two large ¼ scale airplane canopies, The plaster molds were left inside for the photo. These parts filled the full 12 x 18 inch sheet and were 5-6 inches tall. The shape determines the maximum draw more than the machine does.

Years ago I recognized a void that no one was filling. I found a few small vacuum forming machines being produced but they all had poor performance with high price tags. To this day no one sells a "full powered" small machine that will fit into a hobbyists budget. For three years I filled this void by manufacturing a popular line of small machines called Hobby-Vac. This is where I learned about the realities of retail pricing. It turns out that a powerful small machine <u>can</u> be built for a low price, it just can't be <u>sold</u> for a low price. The truth is that every product has to retail for 4 to 5 times what it cost to build. The manufacturer often makes less than the distributors and retailers that sell them. This outrageous profit structure offended me as a consumer and manufacturer. These machines were easy enough for me to build at home so with some instruction you can build one too.

Get "industrial" performance at a "Hobbyist price.

Overview and Specifications

HOBBY-VAC	Specifications:

Plastic Sheet Size   12 x 18 inches
Usable Area   10 x 16 inches
Depth of Draw   Up to 6 in. depending on shape
Max Plastic Thickness   1/4 in. for some plastics
Max Vacuum   Up to 29 IN. HG. depending on pump used
Oven Requirements   115 Volts, 14 Amps
Machine size   21 wide x 38 long x 8 inches high

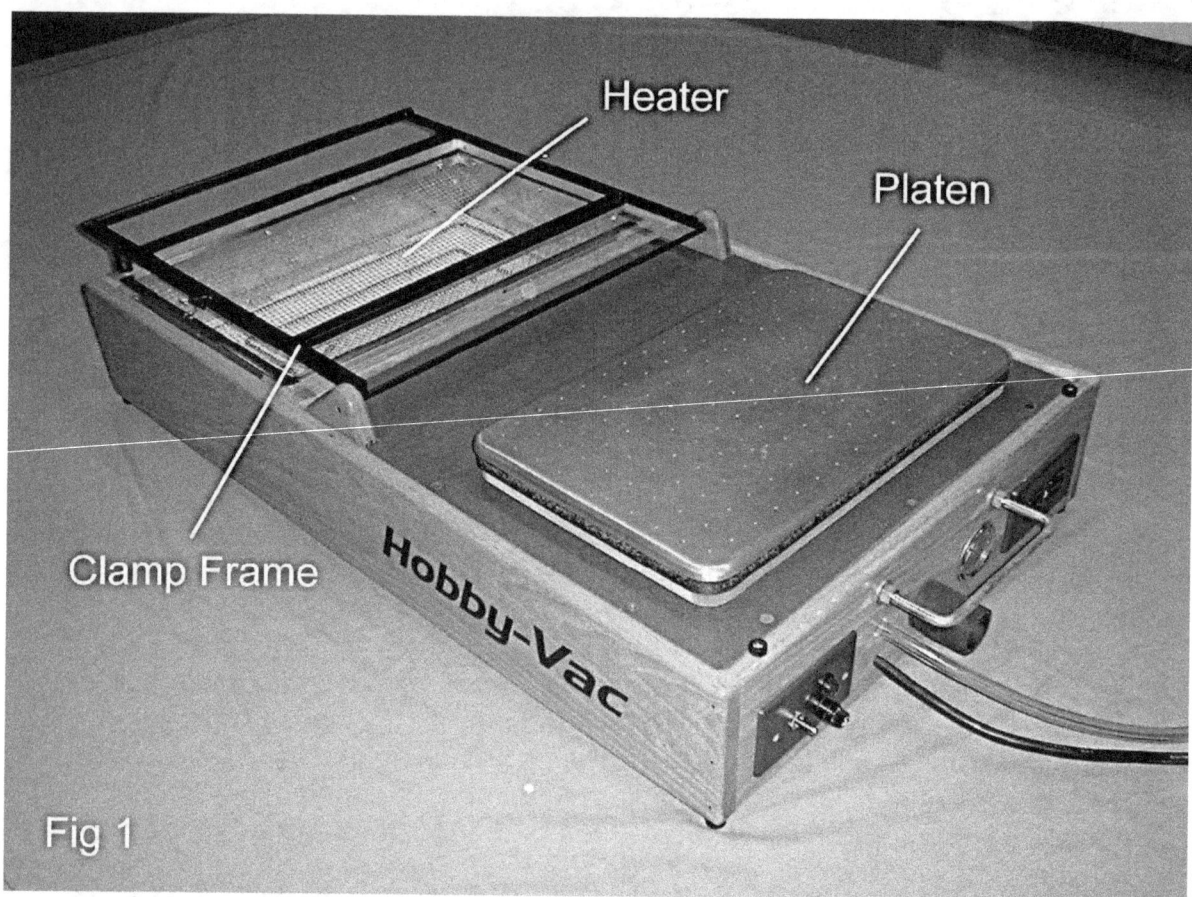

Fig 1

The internet is full of free tutorials showing you how to hook a shop-vac to a wood box and use your oven or rig up some elements from a toaster. Like many others, that's the first thing I tried. It's a great teaser but with very limited forming power and it leaves you wanting more. I needed some real performance for my projects and had used commercial machines before so I knew it could be much better. That sent me into a deep dive of vacuum systems and infra-red heating elements.

This simple Hobby-Vac machine is a real sleeper and can match or outperform commercial machines costing thousands. It looks so simple but all the critical engineering is baked into the design. A tuned infra-red output for the heating. An airtight low volume platen. It's all optimized for use on a 120 volt residential power. Form plastic sheets over simple wood patterns or found objects. Or,.. use your 3D printer for making molds and you have a powerful desktop manufacturing system. **Give it a try!!**

Hobby-Vac Features:

<u>**High Performance Oven -**</u> Uses both convection and infrared radiation and is "tuned" to match the absorption characteristics of plastic sheets. Proven design concentrates heat around the perimeter and in the corners where it's needed. Easy to install heating element kit's are available to cut building time. Controlled airflow around the oven keeps the machine cool even after extended use. Oven operates on a standard 15 Amp. residential house circuit.

<u>**Optimized Forming Surface -**</u> (also called Platen) Airtight sandwich construction with solid wood base. Super low internal volume for fast response. Perforated aluminum top surface for durability and reliable sealing. No gaskets or seals to wear out or replace. Platen is removable, make any size you need. Plans also show a quick build option that is ideal for odd sizes you don't use much.

<u>**Sturdy Wood Cabinet -**</u> Easy to make from standard size lumber. All straight cuts with no fancy joints. Can even be built without power tools. Vented design stays cool when operating.

<u>**Simple Clamp Frame -**</u> Easy to use "Flip Style" clamp frame holds plastic securely with minimal waste. Simple two piece design uses spring clips to hold plastic sheets up to 1/4 inch thick. A "no weld" version is also included.

<u>**Vacuum System -**</u> Plans show several methods to choose from. 1 or 2 stage systems with manual, Air powered or electric pumps. Budget to performance options to fit your needs. Vacuum system is separate and upgradable. Includes sources for low cost pumps.

<u>**Easy to Use -**</u> Short cycle times typically 1 to 3 minutes per part. Fast warm up and no mess to clean up. Uses inexpensive plastic sheets, Optional mobile cart keeps it ready to use.

Hobby-Vac Performance

The performance of your Hobby-Vac machine depends largely on your choice of vacuum components. The vacuum system is separate from the machine and this gives you the flexibility to upgrade it at any time. A later chapter will give you advice on choosing the right components to fit your needs and your budget.

This machine is a little powerhouse, however, you still need to be realistic about what can be vacuum formed. It's important to understand that the shape of your parts is limited by things such as thinning, webbing and mold removal. Your Hobby-Vac has nothing to do with these shape related limitations. The process is always subject to the limitations of the plastic and mold shape.

Some simple rules to follow.

— Your pattern can not have undercuts. These are areas where the plastic can wrap around or under the pattern in such a way that you can't remove the formed part.

— The plastic will shrink as it cools. This shrinkage is in the range of .005 to .008 in. per inch. This can be as much as 1/8 inch. over a 16 inch length. If you can't tolerate this then you must make the pattern oversize.

— You should always provide a slight angle on vertical sides of your pattern. This is called "Draft" angle and should be at least 3 to 5 degrees per side. The reason for this is the shrinkage mentioned above. If you have tapered sides then it will come off more easily.

— The wall thickness will not be uniform on a finished part. The plastic will be the thickest where it first touches the pattern and the last part to touch will be the thinnest. The amount of thinning depends entirely on how much the plastic has to stretch.

— There is a general rule of thumb that the pattern should not be taller than it is wide. This can result in webbing (wrinkles) in the plastic usually coming from a tall corner, as well as excessive thinning. This depends a lot on the shape of the pattern with rounded shapes being better than square ones. The pattern should also be placed at least as far from the edge of the platen as it is tall. If it's too close then the plastic can get thin or not seal to the platen. Be realistic with the size and shape of your mold

<u>Using the Plans</u>

These plans were revised and updated with more info in 2024. I suggest you skim through the plans first so you have a good mental image of the machine while you read this manual. You will notice a lot of helpful **"*Tips*"** and more importantly, **"Cautions"** scattered throughout this manual. The **"*Tips*"** are like little bonus bits of information that might make building easier, or explain why something works. The "**Cautions**" usually deal with safety or a critical step in the building process that can affect performance.

Caution: Read all cautions! See how well that works.

If you have questions about the plans, don't hesitate to reach out for assistance,

Tech Support

hobbyvac@yahoo.com

Chapter 2 - Getting Started

Vacuum forming is unusual because everyone wants or needs a different size machine, there are no standard sizes. I also offer another set of plans that covers larger machines (Proto-Form 2x2, 2x3 and 2x4 ft.sheet sizes). However, this set of plans for the Hobby Vac machine covers only one size. 12 x 18 inches is the maximum sheet size that can be heated with a 120 volt oven operating on a 15 amp. residential house circuit. Now any hobbyist can do vacuum forming in their home.

Of course you can form sheets smaller than 12 x 18 if you make an adapter, but it's not that easy to enlarge this machine without re-thinking almost every part. If you need a larger machine, that's fine and I have other plans that can help, but this manual deals only with the 12 x 18 size. You are welcome to use this information as a starting point to build a custom sized machine, but for liability reasons, I can't offer any engineering assistance.

Keep it Simple

It's just human nature that people love to accessorize, automate and generally add useless features to their machines. I sold a machine to one aerospace company that complained about the price and then installed over $1000 worth of sensors and controls on it. He got mad when I called it "Robo Vac", but it didn't make the parts any better or faster. Here are the top three most unnecessary modifications.

Temperature Control - Yes it is possible to install a simple temperature control from a kitchen oven, but we just don't need it. Kitchen ovens use convection or hot air to heat your food. This machine uses infrared radiation at the proper wavelength for efficient absorption. Don't mess with it, a better indicator of readiness is how much the plastic sags when it gets soft.

Why not use a timer? - As the machine and mold warm up, the cycle times will get shorter. This can happen for 15 or 20 parts before it starts to stabilize, and the next batch of plastic you buy will be different. The best use of a timer is as a safety shutoff.

How about a hinged clamp frame? - Every one imagines a hinged clamp frame that opens like a book and has two latches on it. This machine can form plastics from .005 to .250 thick and the hinges and latches would have to be very adjustable. What you really need is uniform clamping pressure all around the frame, not just on two sides. The cheap little spring clips work so well, I just haven't been able to improve on them. Feel free to embellish the clamp frame with toggle clamps or hinges, it won't affect anything else. However you should resist making changes to the oven and platen because those have a major effect on performance.

Parts Sources

Most of the parts to build this machine can be found locally. The wood can be purchased from the local lumber store and the nuts and bolts found at any hardware store. Electrical parts are also easy to find. The heating element kit's however are made of materials you can't find in stores. So I sell kit's for them and also discuss alternatives in Chapter 8. Sources for vacuum pumps are in Chapter 6, and steel and wood cut lists are included in the plans

Some of the harder to find items are listed below with part numbers from two mail order/online companies. Other parts sources are throughout the book.

McMaster Carr Co. is a huge industrial supplier with the best inventory and service I have found, they are a delight to deal with and have warehouses across the US. You can see their catalog at www.mcmaster.com My second favorite choice is **Grainger** at www.grainger.com

---

Vacuum Gauge - 2 in. panel mount, 0-30 in.hg.

Qty 1, McMaster Carr #4002K24   Grainger #1X501

1/2 in. NPT **1/4 turn Ball Valve** for tank systems only

Qty 1,   McMaster Carr # 33325K23   Grainger # 5X715
          (Use with tank type vacuum systems)

U - Bolts to mount ball valve to machine, 1/4 - 20   Qty 2

McMaster Carr # 3043T644   or Local Hardware Store

1/2 NPT x 1/2 barb **right angle fittings**   Qty 2

McMaster Carr # 53055K192   Or Hardware Store
(use with ball valves mounted on front panel)

3/8 NPT x 1/2 barb **straight fittings**   Qty 1

McMaster Carr # 53055K218   or Hardware Stores
(bottom of platen - for tank systems - use one for each platen size)

3/8 NPT x 3/8 barb **straight fitting**   Qty 1

McMaster Carr # 53055K217   or Hardware Stores
(bottom of platen - for all other vacuum systems - use one for each platen)

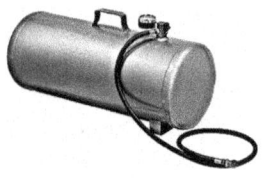

Small **Check Valve**   3/8 to 1/2 ID hose   Qty 1

McMaster Carr # 4610K17   or Hardware Stores
(use between pump and tank - only needed on tank systems)

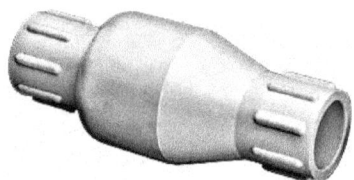

Portable tire filling **air tank** 5 - 10 gallon   Qty 1

Use for tank system only find at hardware or auto parts stores

---

Spring loaded **PVC check valve**   Qty 1

McMaster Carr # 46835K54 or some hardware stores
*(Only needed for a two stage system. Other types may work
as well. These are often used for sump pumps.)*

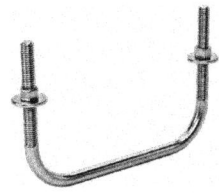

Square U-Bolt, 3/8"-16 Thread Size, 6" Inside Width, 4-5/8" High

McMaster Carr # 3060T51
(Use for lower handle on front of machine)

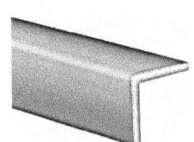

Aluminum angle 1/2 x 1/2 x 1/16 thick

McMaster Carr # 8122A51 or Hardware Stores
(use for top clamp frames and oven trim)

Aluminum top sheet for platen QTY 1

McMaster Carr # 8973K477 or Google for other sources.
(Best material is alloy 3003-H14 or 6061-T3. Thickness is
.040-.060. You may be able to use an aluminum cookie sheet)

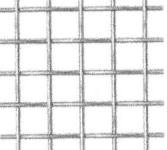

1/4 inch mesh **screen**

McMaster Carr # 9220T426   or hardware stores.
*(Also called Hardware Cloth. use for platens and oven screen - get
enough for extra platens)*

Safety Timer - This adjustable timer will automatically shut off your oven
in 1 to 5 minutes. Just twist the knob to re-start it.5 Minutes is shorter than
a typical heating time so its purpose is to shut off if left unattended.

Made by Intermatic model FF5M

McMaster Carr # 7014K6   Grainger # 6X545

Fuse Holder - panel mount 15 Amp.   Qty 1

McMaster Car # 7087K15   Grainger # 6F457
(*uses 1/4 x 1 1/4 fuses - must be rated at 15 amps or more - Grainger
one is heavier, 30 amp.)*

Fuses 15 Amp. ceramic style   Qty 1
        McMaster Carr # 71385K527
(*Must be ABC ceramic body 15 amp. 250 volt rated - not 32 volt glass tube type*)

Pilot light - Red -120 volt   Qty 1
McMaster Carr # 2779K15 or similar

15 Amp DPDT Toggle switch   Qty 1

Must be ON-OFF-ON type to control both heat and vacuum

McMaster Carr # 7343K821

Chapter 3 - Build a Mobile Cart

This is optional,.. Your Hobby Vac machine will work just fine on a workbench or your kitchen table, but there's nothing like the convenience of having it all hooked up and ready to go at a moment's notice. This mobile cart has room underneath for the vacuum system and an extra shelf to store plastic sheets or accessories. You can also purchase various types of plastic and metal utility carts that may work, but none of them will fit as well and they typically cost well over $100.00. You can build this simple wood cart in a few hours.

The sketches show two views of the cart. The letters in circles correspond to the wood cut list below. Cut all pieces to size and assemble as shown. The 32 inch overall height seems about right, but you may want to adjust it if you are shorter or taller than normal. The shelf placement is also just a suggestion and you may have to leave out the middle shelf if your vacuum components are too large. I highly recommend using swivel casters on all four legs so you can roll it around the shop. Shorten the legs as needed to accommodate the casters.

Wood Cut List for Mobile Cart

Part	Qty.	Size	Material	Description
A	2	34 in.	2x4	Top supports F&R
B	4	31 in.	2x4	Shelf supports F&R
C	6	19 in.	2x4	Side supports
D	4	32	2x4	Legs
---	1	24 x 36	1/2 Plywood	Work surface
---	2	24 x 28	1/2 Plywood	Shelves
---	4	2 in. dia.	Rubber	Swivel Casters

Tips:

-- Use drywall screws instead of nails so everything stays tight.
-- Use 1/2 in. or thicker plywood or particle board for work surface and shelves.
-- You could cut a hole in the top surface for the vacuum hoses and electrical
   cords to pass through to the shelf below.

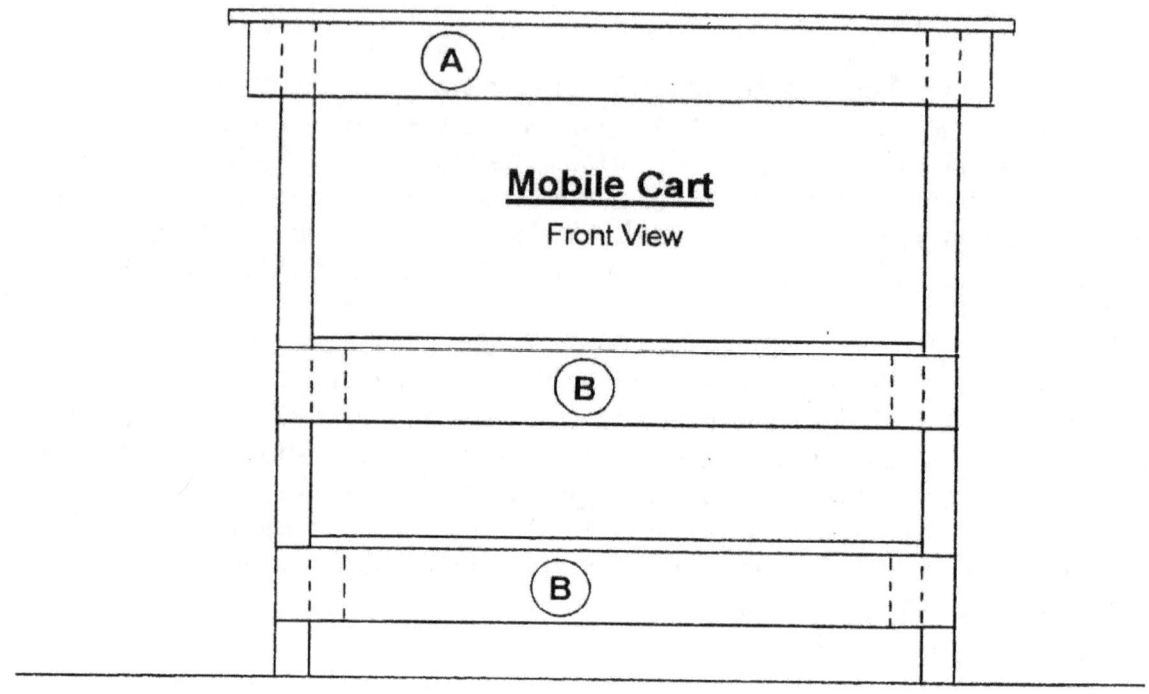

Mobile Cart

Front View

A

B

B

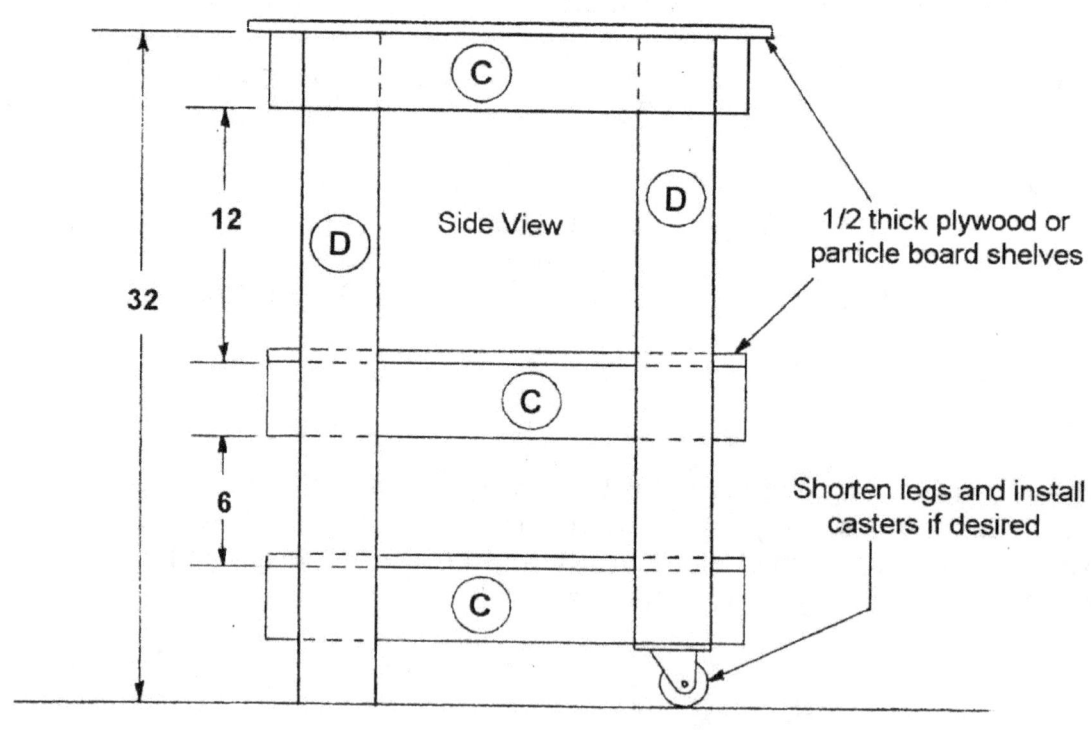

C

32

12

D

Side View

D

6

C

C

1/2 thick plywood or
particle board shelves

Shorten legs and install
casters if desired

Hobby-Vac --- Drawing # 2

Chapter 4 - Building the Wood Cabinet

The wood parts for the machine itself are about as simple as I could make them. I tried to think of the guy who has no power tools, so I only used standard width lumber, straight cuts and no fancy joints. You can join the parts with nails or screws, but I suggest you also use glue on any parts that don't have to be removable. Do not glue or nail the top and bottom covers or the pivot brackets.

A table saw is the best choice for straight cuts and identical matching parts, but they can be done carefully by hand, just make sure you assemble it on a flat surface so you don't build a twist into it.

Tips:

-- You will have to pre-drill all screw holes in hardwood, and even a clearance hole for the top layer so the screw can pull them tightly together. Practice on some scraps until you find the right drill sizes so the screw goes in fairly easily, yet holds securely. You can use a little soap on the screws for lubrication. -- It is possible to use nails in hardwood if you pre-drill those holes too, or use a pneumatic nail gun.

Materials:

Cabinet:

The cabinet itself can be made from any type of wood, but if you have access to power tools, it's not much more trouble to make it from a hardwood such as Oak or Maple and have something worth leaving to your grandchildren. If you have to make all of the cuts by hand, then I suggest a softer wood such as pine or poplar. The top cover can be made of plywood or particle board.

Bottom Cover:

The bottom cover under the oven is made from pegboard for ventilation. We will still have to enlarge some of the holes to get enough airflow through and around the oven. The holes should be no larger than 1/2 in. dia. so you can't stick a finger through and touch the electrical parts. Refer to the drawings and photos for the hole locations.

Top Cover:

The top cover of the machine is where the Platen mounts and can be made from 1/2 inch thick particle board or plywood. I made mine out of something called Medium Density Fiberboard (MDF). which is just like a thick pegboard with no holes. I just like the brown color when it's varnished. MDF or particle board is usually flatter and smoother than plywood. Whatever material you choose can also be used to make the spacers under the platen.

Platen:

The ideal material for the platen is 3/4 inch particle board with a laminate on top. This laminate, also called Formica is what's used on many kitchen countertops and it's crucial to the operation because it forms the inner layer of an airtight sandwich construction. This construction will be covered in more detail in a later chapter, I just mentioned it here so you can get the right materials.

The good news is that anyone who sells or installs kitchen countertops will almost certainly give you this material for free. Just ask for the scraps they cut out to make sink holes, they will probably try to give you a hundred of them. make sure to get enough for any reducers you want to make

Tip:

-- There are three common types of laminate. The one you want is the thickest (almost 1/16 in.) and is always used in flat sheets because it's too thick to bend. The second type is much thinner and is used on the countertops that have a curved backsplash and front edge, avoid this type. The worst choice is called "Melamine" and is often sold as shelving boards or as part of cheap cabinets. This laminate is almost paper thin and is unsuitable for our use.

<u>Overview of Assembly</u>

See the cut-away drawing below for an overview of where things go and refer to it throughout the project. You can see how the oven box is suspended inside the cabinet with an airspace to allow cooling air in the bottom and let it exit through the vent holes around the oven box. See how the clamp frame which holds the plastic can pivot or swing from the oven side over to the platen side. The clamp frame fit's around the platen where your mold will sit. The photographs that follow also show the layout and how everything fits inside a simple wood cabinet.

Refer to the wood cut list to make the individual pieces and try to make all of the cuts nice and square. I suggest you build the cabinet first and then make the top cover to fit inside it. Don't mount the platen or its spacer to the top cover until you have read that chapter, those parts need to be aligned with the clamp frame.

If you are building the deluxe platen (with laminate top), it has a 1/4 radius corner around the perimeter that is fairly important. It can be approximated with a file and sandpaper, but the best way is to use a router and a corner rounding bit. You can also use the same router bit to round off all the exterior corners for a more professional look as you can see in the photo's.

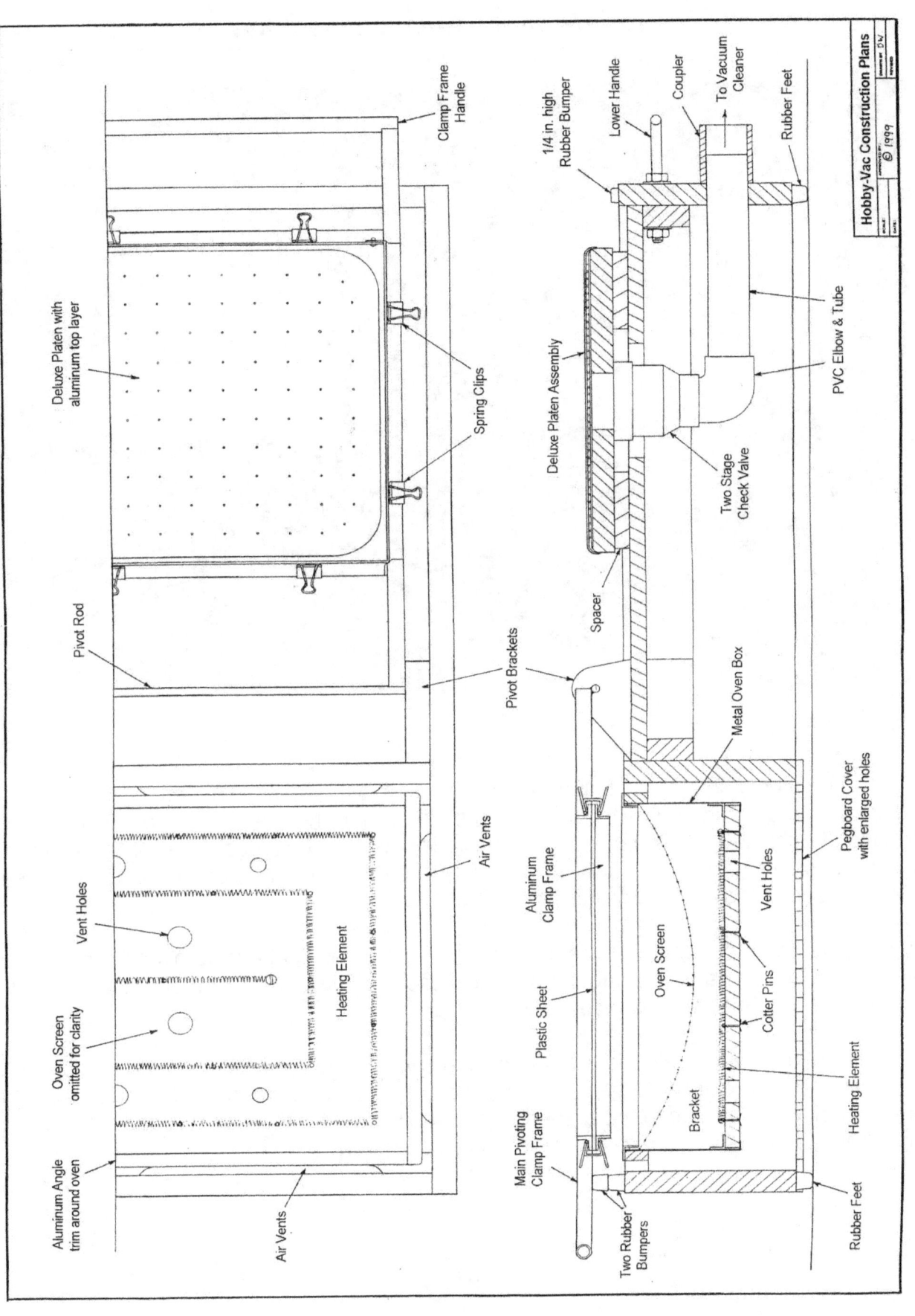

Deluxe Platen with aluminum top layer

Clamp Frame Handle

Spring Clips

Pivot Rod

Air Vents

Aluminum Angle trim around oven

Oven Screen omitted for clarity

Vent Holes

Heating Element

Air Vents

1/4 in. high Rubber Bumper

Lower Handle

Coupler

To Vacuum Cleaner

Rubber Feet

Deluxe Platen Assembly

Spacer

Pivot Brackets

Two Stage Check Valve

PVC Elbow & Tube

Metal Oven Box

Pegboard Cover with enlarged holes

Main Pivoting Clamp Frame

Aluminum Clamp Frame

Plastic Sheet

Oven Screen

Bracket

Cotter Pins

Vent Holes

Heating Element

Two Rubber Bumpers

Rubber Feet

Hobby-Vac Construction Plans

© 1999

Top View of a finished machine

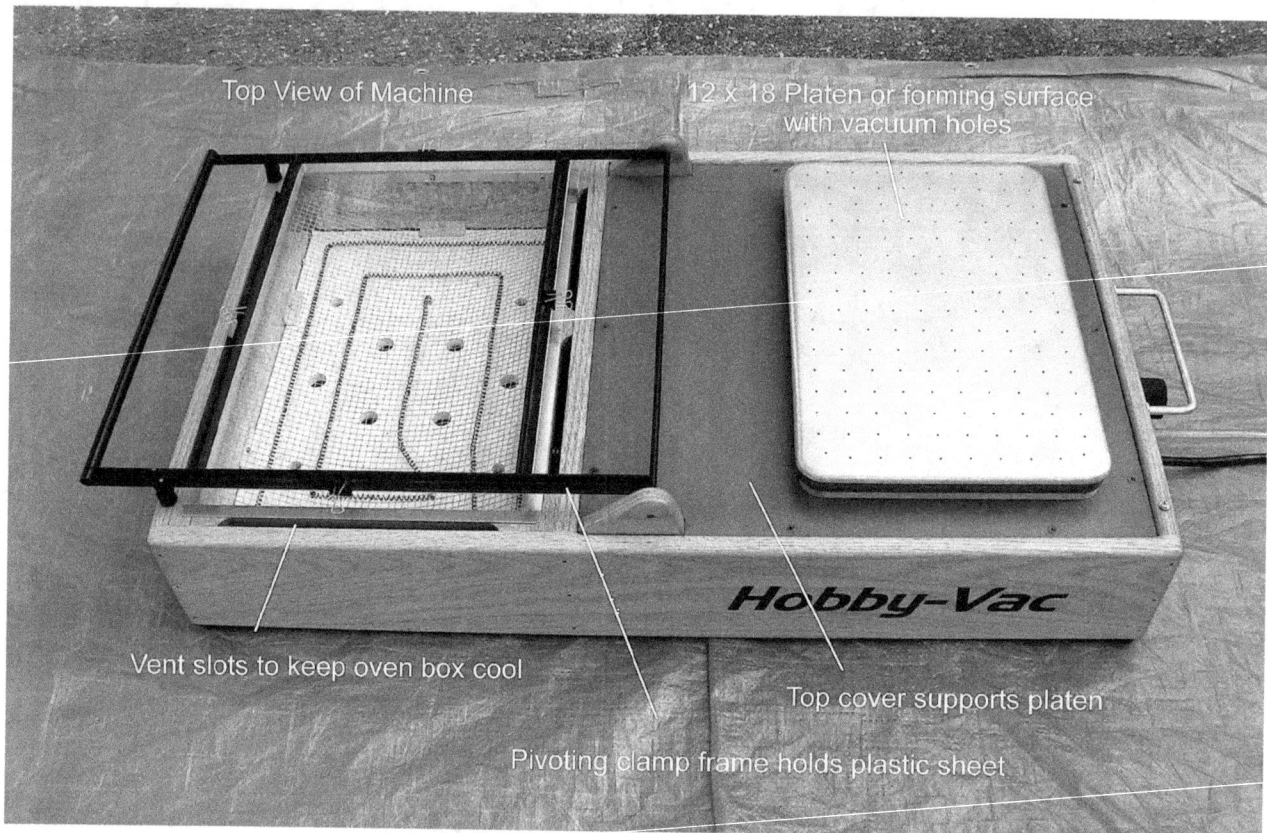

Top View of Machine

12 x 18 Platen or forming surface with vacuum holes

Vent slots to keep oven box cool

Hobby-Vac

Top cover supports platen

Pivoting clamp frame holds plastic sheet

Top view photo of the finished machine is shown above.. The left side houses the heating element and you can see the vent slots around the top of the oven. This keeps the wood cabinet cool. The pivoting clamp frame holds the plastic sheet and swings from one side to the other. The forming surface is on the right,.. note the vacuum holes You can see that there's very little inside.

It's hard to see but there is a screen with ¼ square holes suspended over the heating element. This will catch the sagging plastic if you let it go too far. You can also see this screen in the large cutaway side view drawing

The platen on the right side is held down with two screws up from the bottom and can be changed for a different size. The black clamp frame is made from ½ x ½ steel angle and is welded or brazed in the corners. The plastic sheet will clip to this with spring clips so it can flip from the oven to platen when ready to form.

Bottom view of finished machine

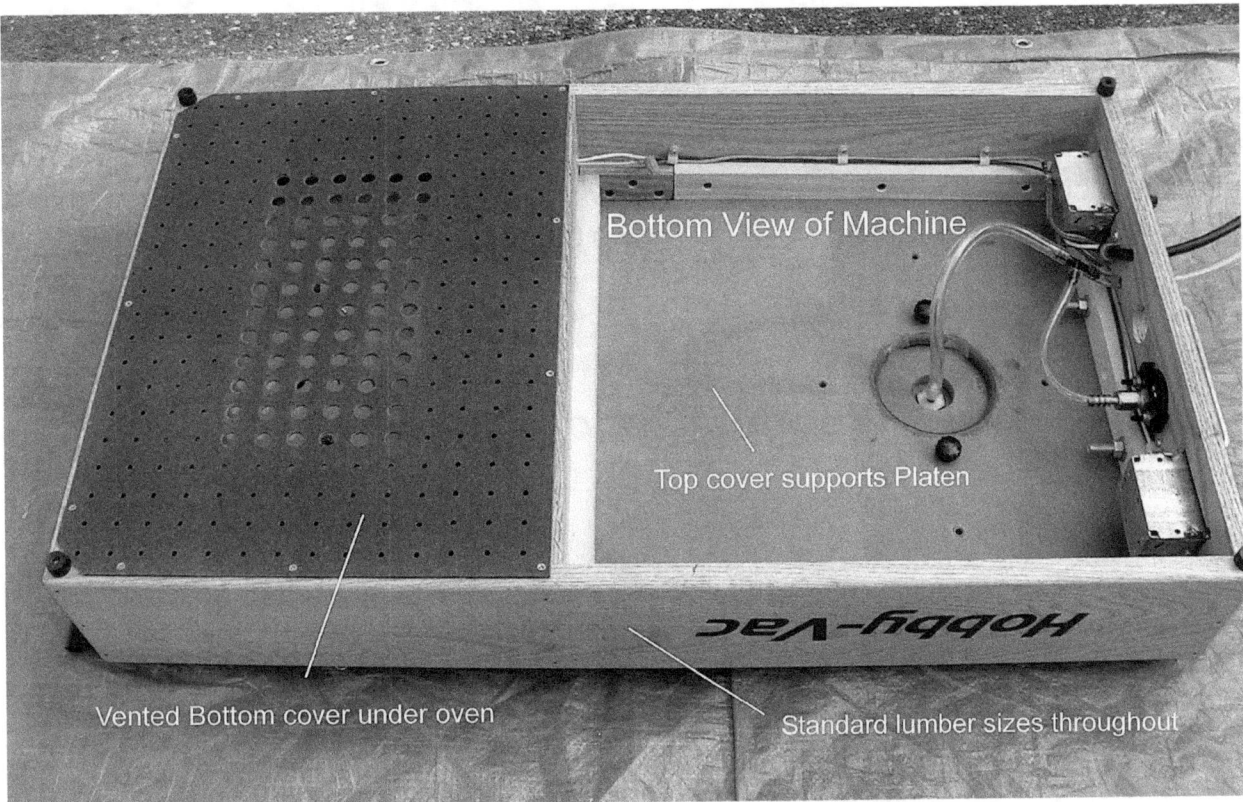

Bottom View of Machine

Top cover supports Platen

Hobby-Vac

Vented Bottom cover under oven

Standard lumber sizes throughout

This is a view into the bottom of the machine, you can see there is very little inside. On the right you can see standard electrical boxes and some wiring that leads to the heating element. This is an early machine and later chapters show three electrical boxes instead. The placement of all components is flexible so just make it fit your components,you have plenty of space to work in. The metal boxes will be grounded so nothing underneath can be a shock hazard.

The two black round objects are plastic thumb screws that fasten the platen in place on the top side. A smaller platen is installed in that photo. The clear vacuum hose attaches to the smaller platens. It is currently using a direct pump system without the two stage valve.The large access hole in the top cover allows room for a two way check valve if you are running the two stage vacuum system. Other photos in these plans will show that valve in place.

You can see the brown pegboard bottom cover over the oven side has some of the holes enlarged for better airflow. Air will enter there and exit through vents on top to keep the cabinet sides cool.

Wood Cut List

On all the prior drawings you will notice letters inside circles for each wood part of the cabinet. Those letters correspond to the chart below giving dimensions. These are all standard size lumber with square end cuts for simplicity. If you cut all the boards to length then assemble according to the drawings everything will fit. Please note that the advertised sizes of lumber does not match the actual size. For example a 1 x 6 board actually measures ¾ x 5 ½.

Wood Parts for Cabinet

Part	Qty.	Size	Material	Description
A	2	5 1/2 x 35 x 3/4 thk.	1 x 6 lumber	Sides of cabinet
B	3	5 1/2 x 19 1/2 x 3/4 thk.	1 x 6 lumber	Cabinet F-R & mid.
C	2	1 1/2 x 15 x 3/4 thk.	1 x 2 lumber	Top support - Side
D	1	1 1/2 x 19 1/2 x 3/4 thk.	1 x 2 lumber	Top support - Front
E	1	1 1/2 x 18 x 3/4 thk.	1 x 2 lumber	Top support - Rear
F	2	3/4 x 19 1/2	3/4 square	F&R vent strips
G	2	3/4 x 12	3/4 square	Side vent strips
H	1	19 1/2 x 19 1/4	1/2 in. thk. sheet	Top Cover
I	1	21 x 15	1/8 Pegboard	Bottom oven cover
J	2	3 1/2 x 3 7/8 x 3/4 thk.	1 x 6 lumber	Pivot brackets

Wood Parts for Platens

Platen Size	Qty.	Top Board	Bottom Spacer
12 x 18	1	10 1/2 x 16 1/2	10 1/4 x 16 1/4
9 x 12	1	7 1/2 x 10 1/2	7 1/4 x 10 1/4
6 x 9	1	4 1/2 x 7 ½	4 1/4 x 7 1/4

Top View of Wood Cabinet

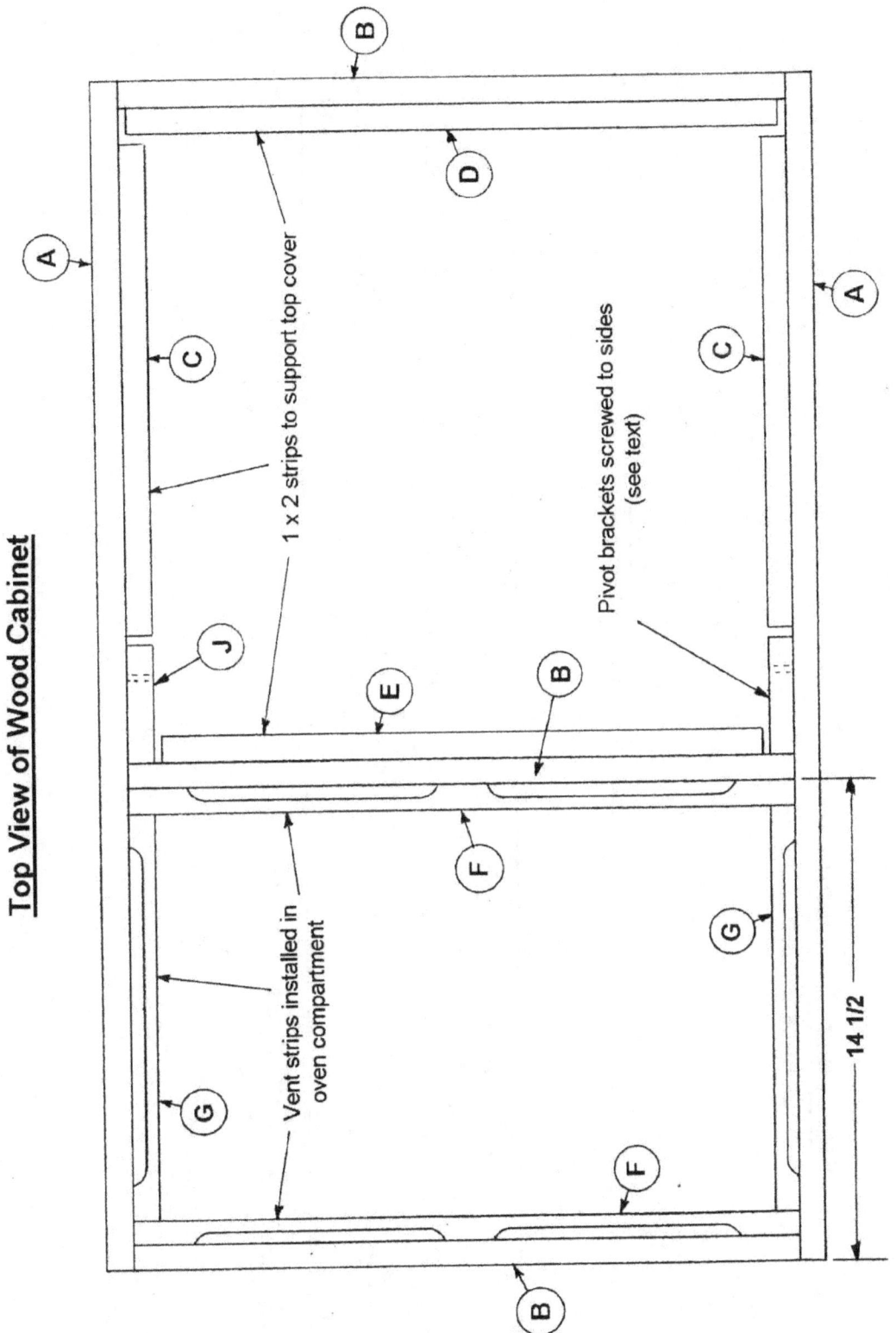

B

A

A

D

C

C

1 x 2 strips to support top cover

Pivot brackets screwed to sides (see text)

J

E

B

B

F

Vent strips installed in oven compartment

G

G

F

14 1/2

Side View Cut-Away
of Wood Cabinet

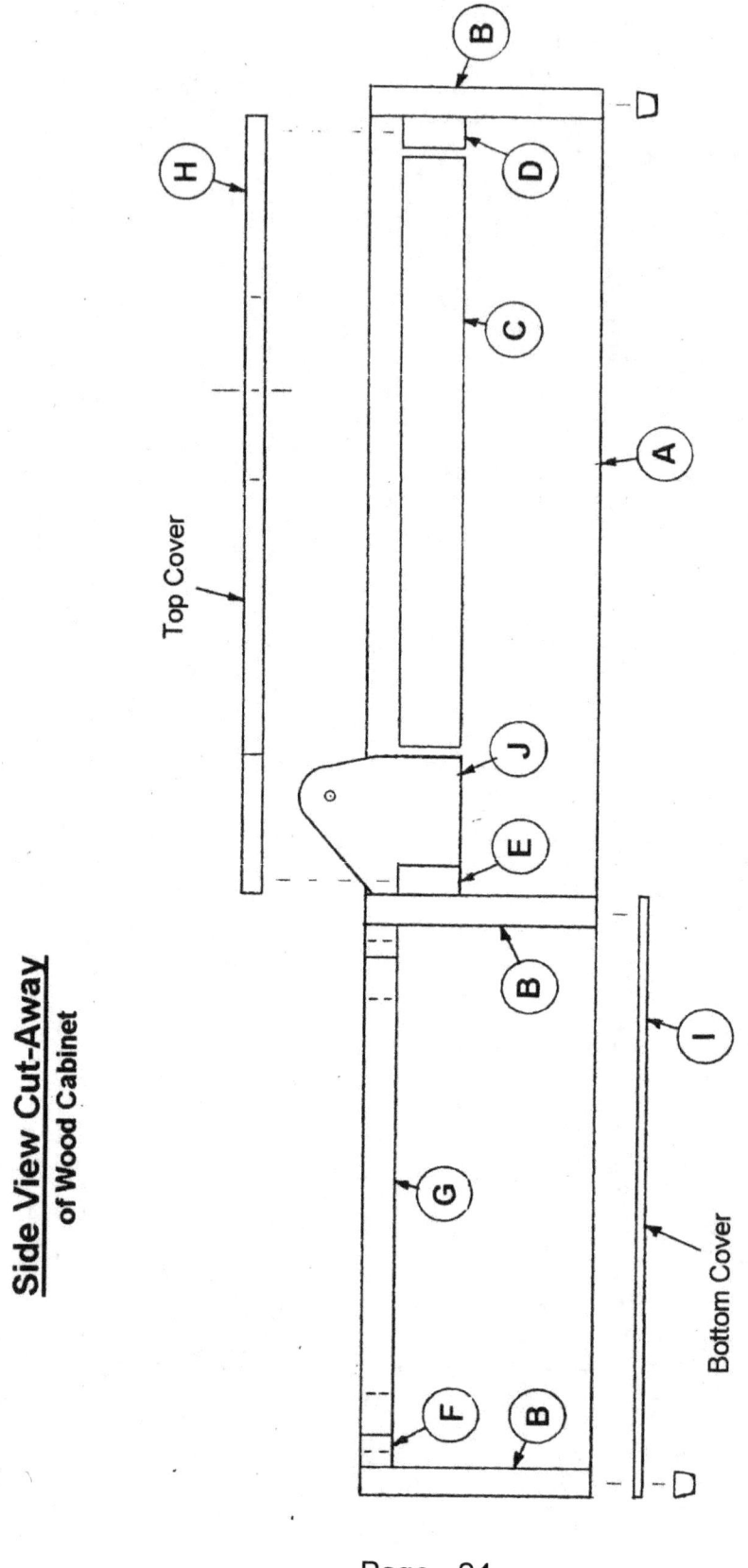

Top Cover

Bottom Cover

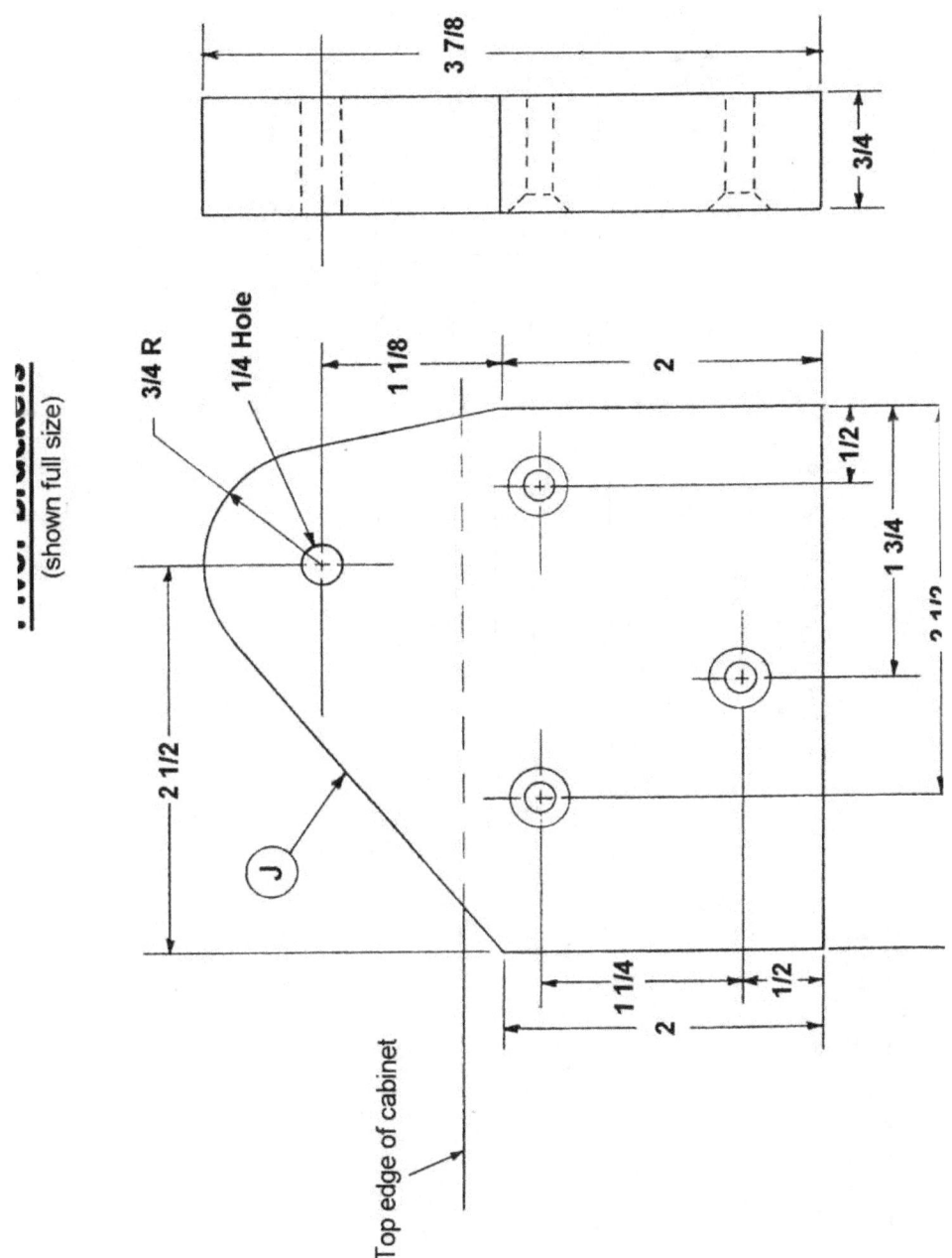

PIVOT BRACKET
(shown full size)

3/4 R

1/4 Hole

3 7/8

3/4

1 1/8

2

1/2

1 3/4

2 1/2

2 1/2

J

1 1/4

1/2

2

Top edge of cabinet

The above drawing shows the pivot bracket, You will need a matched pair of these. The dashed line shows where the top edge of the cabinet is when fully installed.. Make sure the pivot hole is 7/8 above the top of the cabinet and the top cover on the platen side is 1/4 in. below the top edge. This assures the proper relationship between the clamp frame and platen when those parts are installed. The platen should protrude 3/8 inch above the clamp frame surface if you have everything right. The drawing below shows this relationship.

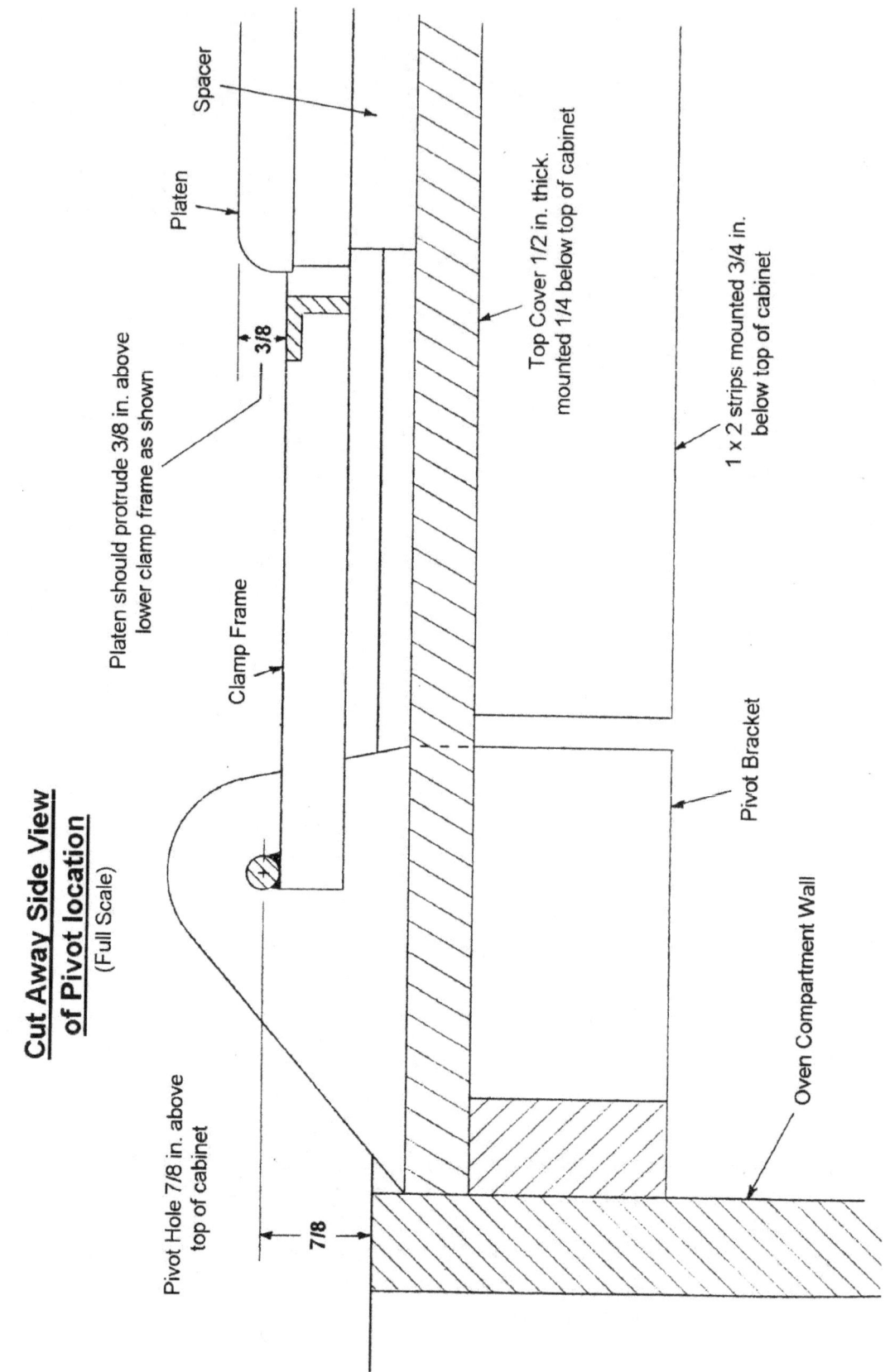

Cut Away Side View of Pivot location
(Full Scale)

Spacer

Platen

Platen should protrude 3/8 in. above lower clamp frame as shown

3/8

Clamp Frame

Top Cover 1/2 in. thick. mounted 1/4 below top of cabinet

1 x 2 strips mounted 3/4 in. below top of cabinet

Pivot Bracket

Pivot Hole 7/8 in. above top of cabinet

7/8

Oven Compartment Wall

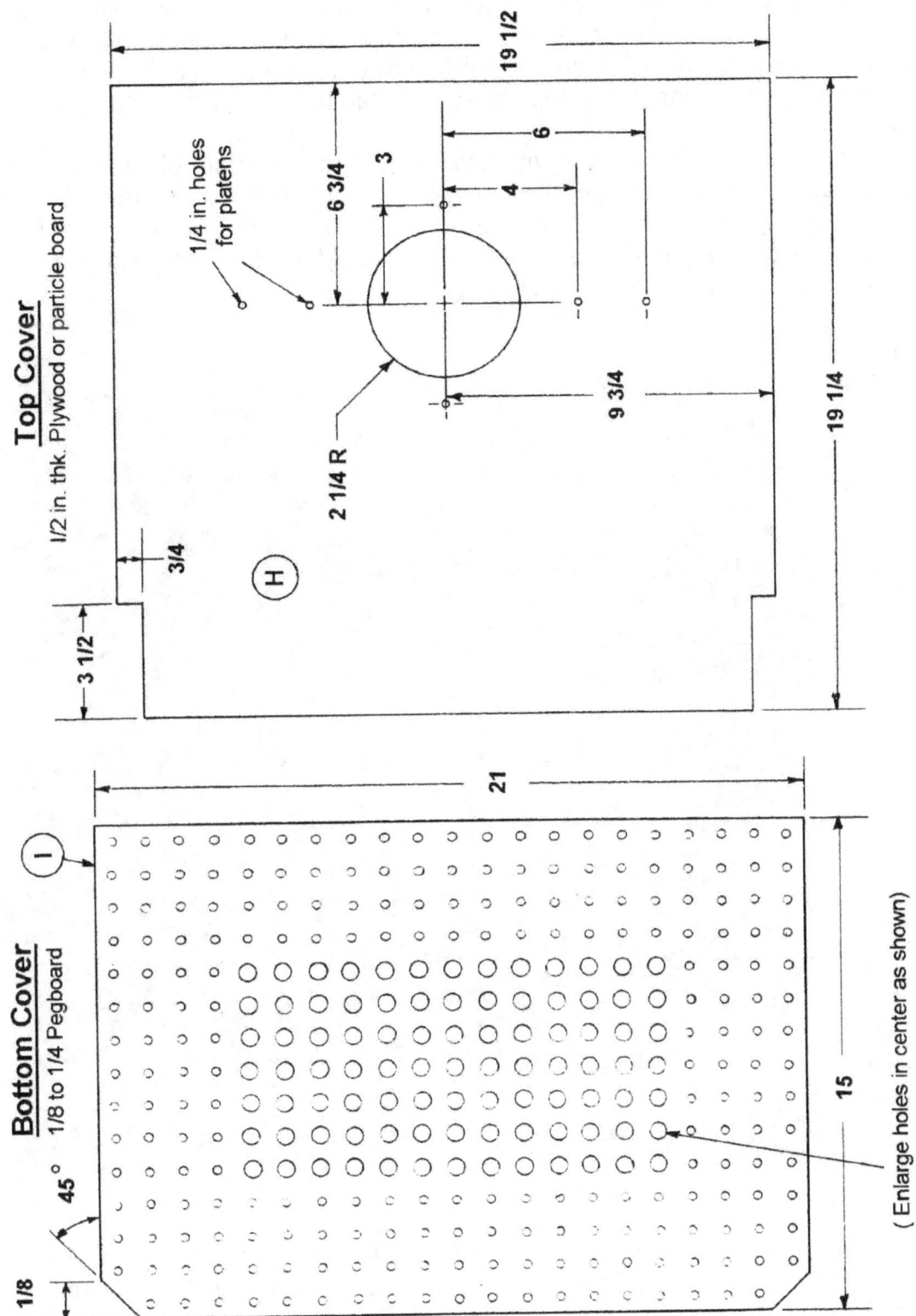

Top Cover
1/2 in. thk. Plywood or particle board

19 1/2

19 1/4

6 3/4

3

6

4

9 3/4

1/4 in. holes for platens

2 1/4 R

3/4

3 1/2

H

Bottom Cover
1/8 to 1/4 Pegboard

21

15

45°

1 1/8

I

(Enlarge holes in center as shown)

The drawing on page 23, is a top view of the wood cabinet with the top and bottom covers removed. The left side is the oven compartment with the vent strips installed. The sheet metal oven box will fit into this opening. The right side shows the 1x2 strips that will support the top cover and also the pivot brackets.

Install the pivot brackets with screws so that the pivot hole is in the proper location as shown on page 26, then install the 1x2 strips so they are 3/4 in. down from the top edge of the cabinet. You can use a scrap piece of 3/4 wood for a depth gauge. These strips can be glued and nailed in place, but the pivot brackets and top cover must be installed with screws for removal later.

Wood strips cutaway then screwed to inside

Vent slots

The photo above shows how the 3/4 x 3/4 wood vent strips have been cut away so when they are installed around the oven area, the space will form slots or vents. Air will rise out of these vents and keep the exterior box cool. I used a corner rounding bit on my router to round all edges for a professional look. This machine was made from red oak with a satin clear coat.

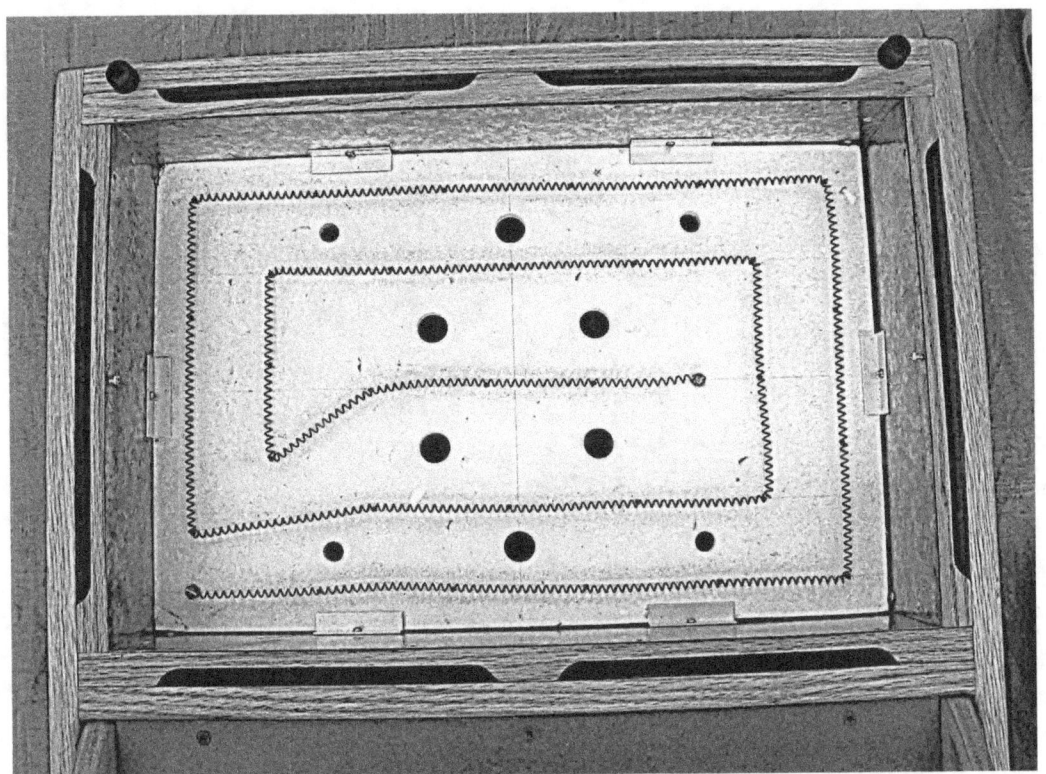

The above photo shows the complete oven enclosure with sheet metal sides and heating element on the bottom. This oven box is screwed to the vent strips with small screws around its upper edge. You can see two screws holding it in place until the screen and optional aluminum trim are installed later.

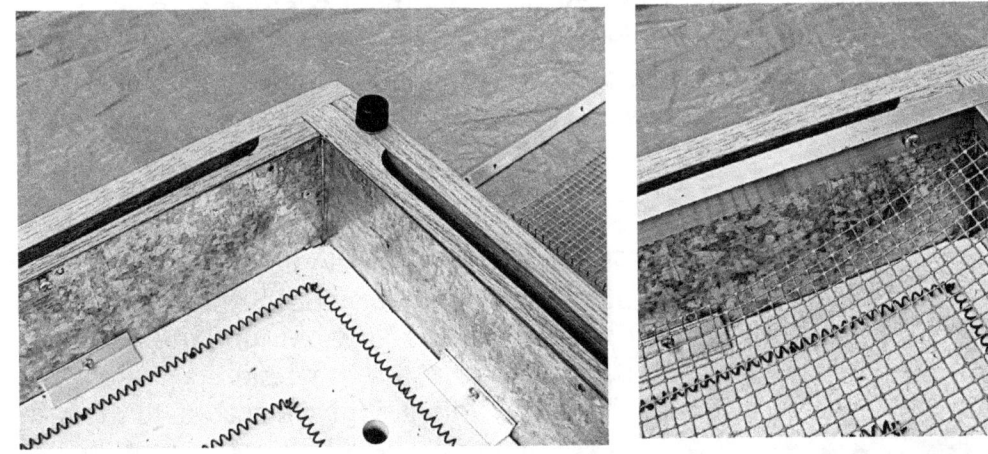

The right photo shows the aluminum angle trim pieces holding the screen and sheet metal box in place. The metal oven box is suspended from these screws.

The above photo shows the wood pivot bracket area and 1x2 strips screwed inside the cabinet. The top cover will rest on these 1x2's and screw in place. You can see how the clamp frame rod fits into the drilled hole in the pivot blocks so it's free to rotate. I cut some thin plastic washers to keep the frame from rubbing on the wood.

The top cover is made from 1/2 in. thick particle board or plywood. Attach it to the 1x2 strips with screws around the perimeter so it is removable
The large hole just gives access to the platen fittings. The small holes near the center are for fastening the different size platens.

When the metal clamp frame is swung over the oven for heating it needs to rest on something to hold it level. You could use a small wood block, but I used two rubber bumpers as shown in the picture below

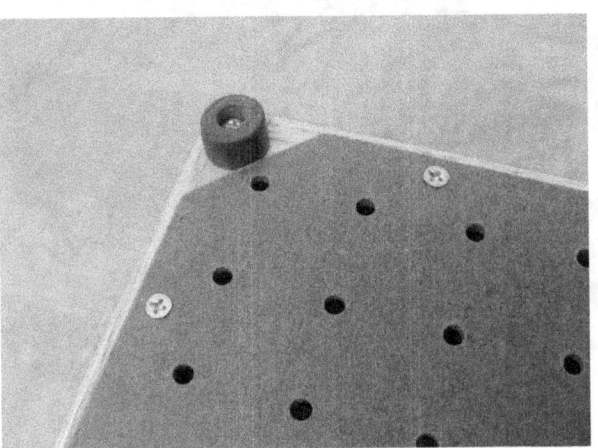

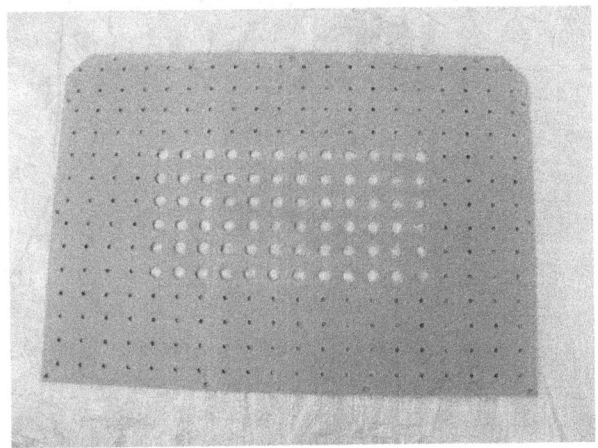

The oven bottom cover is made of pegboard with the holes in the center (6x12 area) enlarged to 1/2 in. for improved airflow. The pegboard can be 1/8 to 1/4 in. thick and should be fastened in place with screws. The corner notches should clear the rubber feet on the cabinet..

Caution!

-- *Do not use a solid bottom cover, there must be enough airflow around the oven for safety.*
-- *Do not leave the bottom cover off or the electrical connections will be exposed to accidental contact.*
-- *Make sure you install rubber feet as shown to raise the machine up and allow clearance for airflow. Do not operate the machine on carpet or block airflow in any other way.*

Chapter 5 - Clamp Frames

The photo above shows the welded steel pivoting clamp frame painted black. It is shown in the heating position with no plastic sheet in place. The handle is on the right and it "flips" in an arc to the left when the plastic is soft.

We'll start by showing the welded version. If you don't have the ability to weld, skip ahead to the "No Weld" version

The Hobby-Vac machine uses a simple pivoting clamp frame that holds the plastic sheet and moves it from oven to platen. It's a "Flip Frame" design which means that it swings over to the oven side for heating and then back where the forming takes place. This is the simplest, most direct way to hold and transport the plastic sheet.

It's true that a straight down motion handles really deep draws better, and this is the system used in my other plans for the industrial Proto-Form machines. But, the flip frame still works great and can accommodate some shapes up to six inches tall on this machine with no problems. While it is true that the plastic is fanned through the air, it only takes about a half second and has minimal cooling effect.

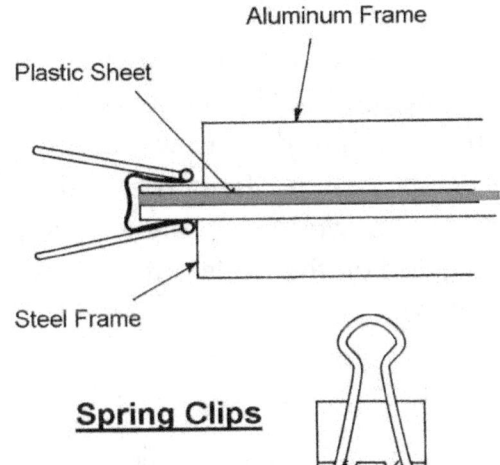

Aluminum Frame

Plastic Sheet

Steel Frame

Spring Clips

There are actually two frames held together with spring clips and your plastic sheet goes between them like a sandwich. The main frame is permanently mounted to your machine on a pivot. The top frame is loose or removable and is made from thinner 1/16 in thick, 1/2 inch aluminum angle. I chose this material because it is lightweight and it warms up quicker than steel. The spring clips for holding the frames together can be found at any office supply store.

The photo below shows the clamp frame pivot, the steel main frame painted black, and the top removable frame is silver colored. You can also see its relationship to the machine. When flipped over the platen side, it actually fits around the platen and travels past the top surface to guarantee contact and a good seal to the plastic. The spring clips are not shown in this picture.

Optional Lower Handle

This is a nice feature you can add to the front of the machine and it only costs a couple of dollars. This handle is actually a large square shaped U-bolt, and it just bolts through the wood box. When you lower the clamp frame, your hand can now grab both handles and squeeze together gently. This holds the plastic against the platen until it cools. You don't need this extra handle, but it just feels natural when it's there and you can actually use it to carry your machine. This U-bolt is listed on the parts but you may also find one in a hardware store.

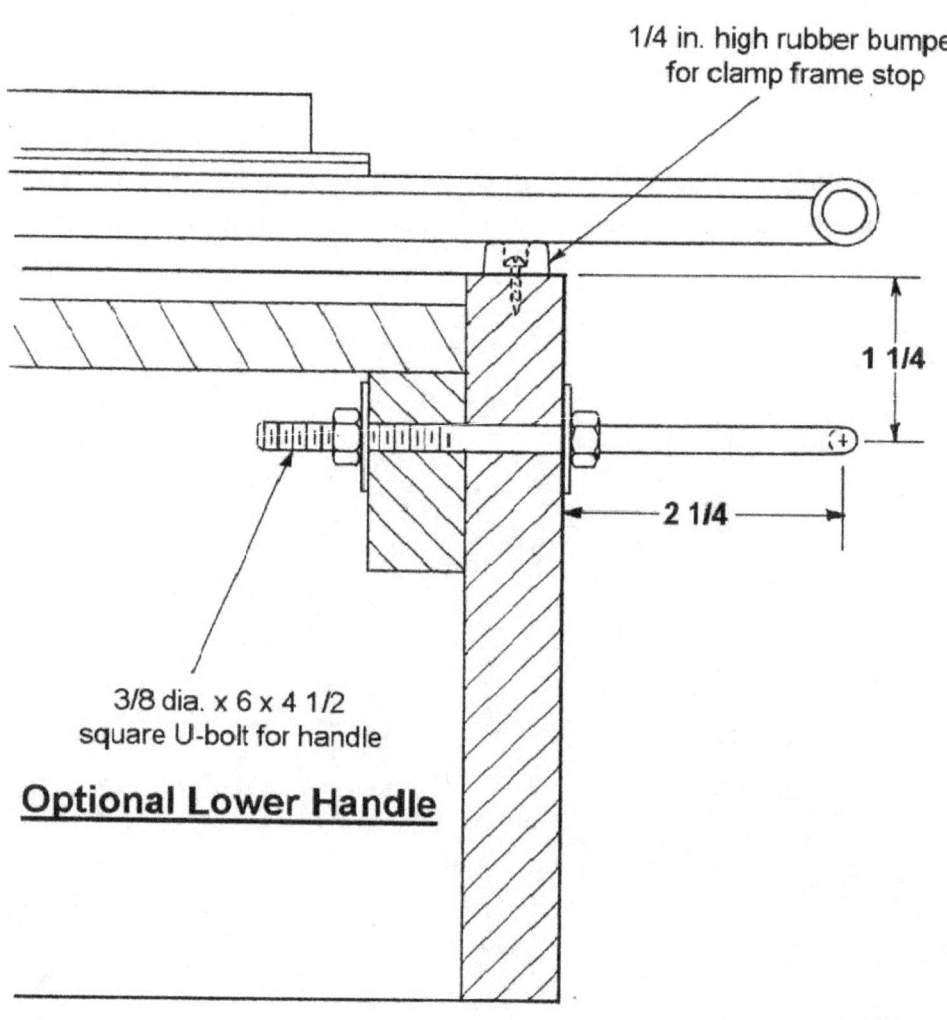

1/4 in. high rubber bumper for clamp frame stop

1 1/4

2 1/4

3/8 dia. x 6 x 4 1/2 square U-bolt for handle

Optional Lower Handle

Welded Clamp Frame Construction:

The pivoting clamp frame stays on the machine all the time and will accommodate the full sheet size of 12 x 18 inches. Most of the photo's show a steel main frame painted black and described below.

If you can't weld or don't want to, skip ahead to the end of this chapter and I will give you an alternative:way to make one that requires no welding. The welded frame is a little more robust but they both work great. If you have welding capabilities you can use 1/2in. x1/2in. steel angle that can be found in many hardware stores or building supply centers, or if you go to a welding shop to get this part made, they can get it for you. The handle shown is actually made from 1/2 in. O.D. steel tubing, but it can be made from square tubing or even 1/2 in. solid rod. It just functions as a handle, but a round shape feels best.

Make sure to weld it together on a flat surface and make sure it's square. If it does warp during welding, you can usually adjust it by cold bending. Make sure the top surface where the plastic sheet lays is flat with no protruding welds or splatter. Below is a side view with end details of the steel frame.

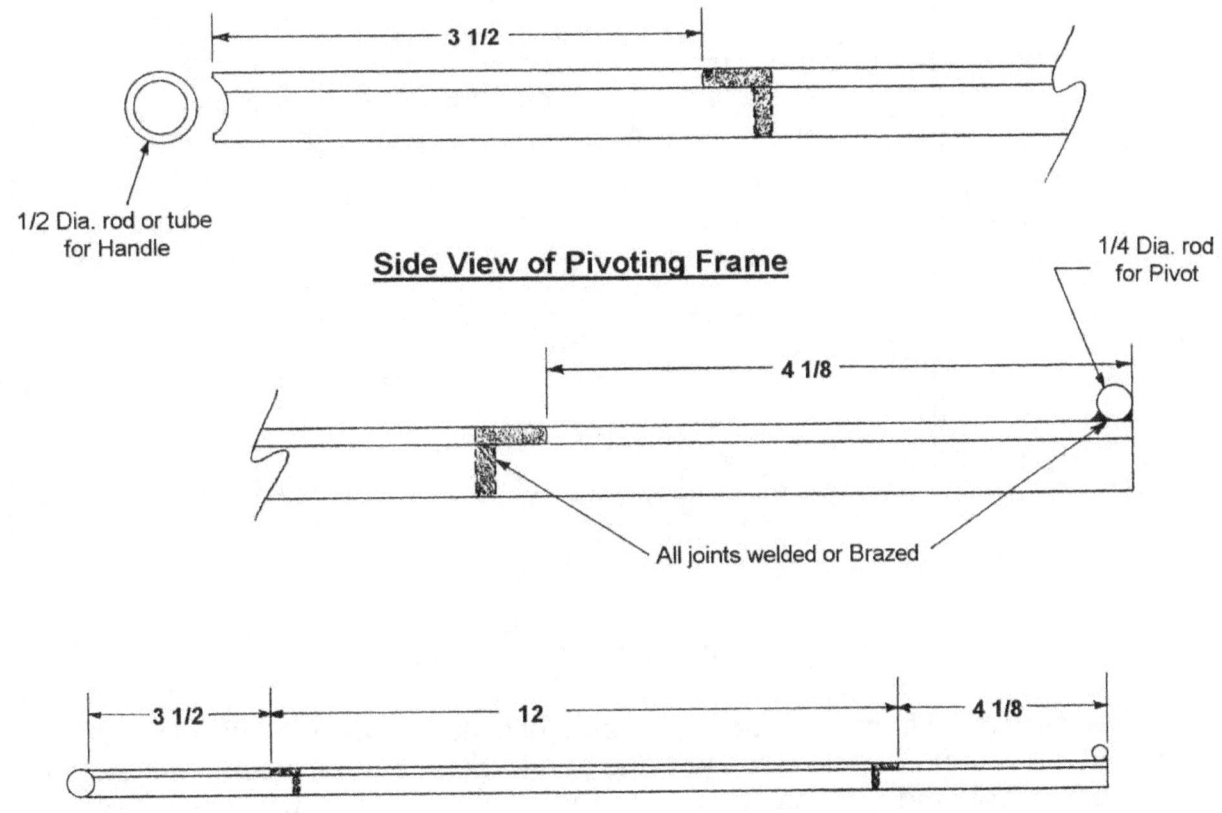

1/2 Dia. rod or tube for Handle

1/4 Dia. rod for Pivot

Side View of Pivoting Frame

All joints welded or Brazed

Below is a top view of the welded steel frame. The right side with protruding ¼ rod is the actual pivot and the ends fit into the wood pivot blocks. Welding or brazing will work equally well.

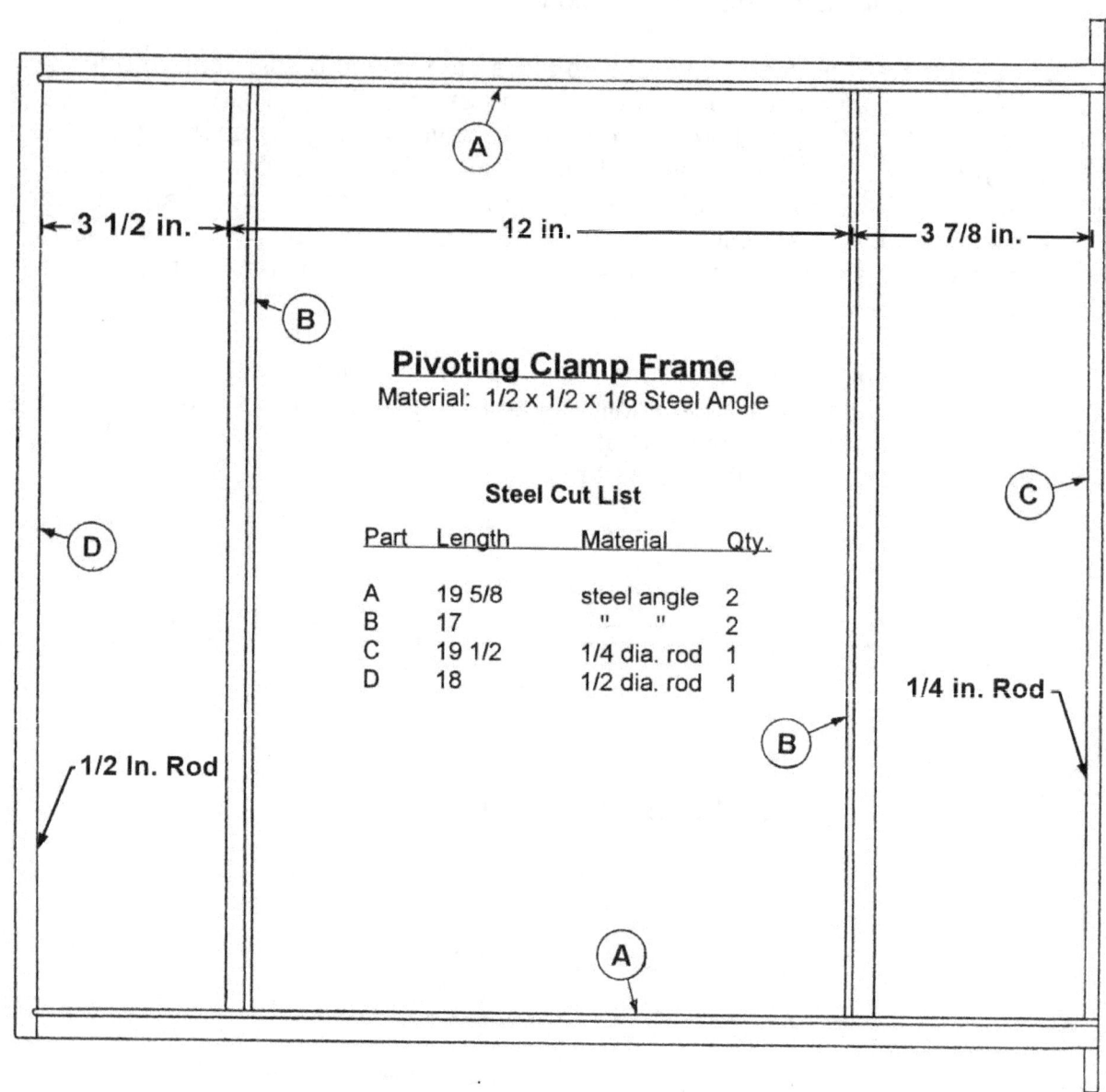

← 3 1/2 in. →   ← 12 in. →   ← 3 7/8 in. →

Pivoting Clamp Frame
Material: 1/2 x 1/2 x 1/8 Steel Angle

Steel Cut List

Part	Length	Material	Qty.
A	19 5/8	steel angle	2
B	17	" "	2
C	19 1/2	1/4 dia. rod	1
D	18	1/2 dia. rod	1

1/2 In. Rod

1/4 in. Rod

Top Aluminum Clamp Frames -

The top clamp frame is a loose part and just sits on top of the plastic sheet to keep it flat. Of course it's held down with the spring clips. You will need one 12x18 inch top frame which fits the main pivot frame and uses full sheets. You will then need another top frame for each size of reducer you decide to make.

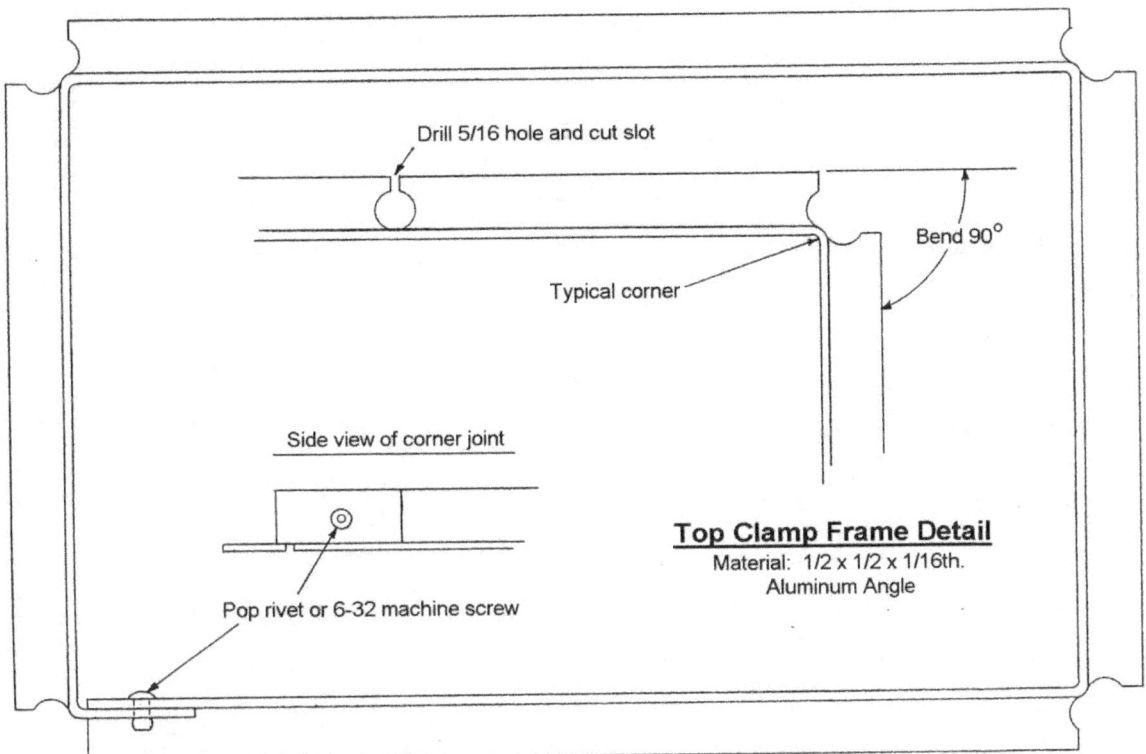

Drill 5/16 hole and cut slot

Bend 90°

Typical corner

Side view of corner joint

Top Clamp Frame Detail
Material: 1/2 x 1/2 x 1/16th.
Aluminum Angle

Pop rivet or 6-32 machine screw

The 1/16 thick 1/2 x 1/2 aluminum angle used for this is commonly sold in hardware stores. This is the same material we will use as trim around the oven opening and you should also get some extra if you plan to make sheet size reducer frames. The top frame is made from one piece that is bent into a rectangle and riveted in one corner. On the next page, I included dimensions for three popular sizes, 12 x18 (full sheet), 9 x12 (half sheet) and 6x9 (quarter sheet).

Note that the bottom edge of the bent tab must be trimmed to sit on top of the flange and keep the top edges level.

Trim bottom edge 1/16th. inch to allow for flange thickness

Upper Clamp Frame Dimensions

Material: 1/2 x 1/2 x 1/16th. aluminum angle

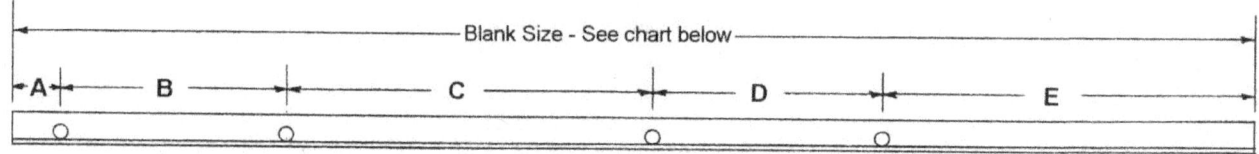

Notch Locations for Top Clamp Frames

Dimension	Frame Size 6 x 9	9 x 12	12 x 18
A	1	1	1
B	5 1/16	8 1/16	11 1/16
C	8	11	17
D	5	8	11
E	7 15/16	10 15/16	16 15/16
Total Length	27 in.	39 in.	57 in.

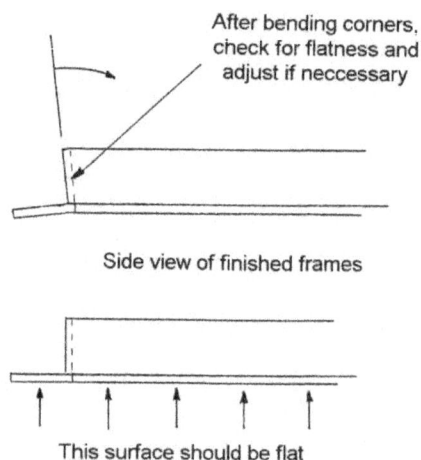

After bending corners, check for flatness and adjust if neccessary

Side view of finished frames

This surface should be flat

Spring Clips

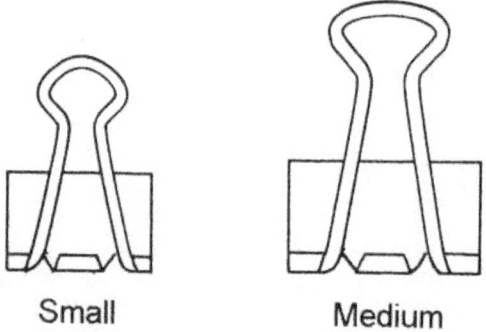

Small   Medium

(shown full scale)

Spring clips work very well because they hold constant tension on any thickness. If using plastic over 1/8th. Inch thick, you can use the medium size clips. I tried every type I could find and the cheapest ones worked the best. These are commonly called "Binder Clips" and are sold in most office supply stores. They come in many sizes, but we can use the small or medium ones. The small clips fit better on the reducer frames.

If you have trouble with the plastic sheet pulling out when forming, check to see if the frames are flat and mate well together with no gaps. Another possible cause would be trying to form the plastic before it is soft enough. You can add more clips or as a last resort you can glue sandpaper strips to the frame for more friction.

"No Weld" Clamp Frame Option

As I mentioned at the top of the chapter, I am providing an alternate method to build the pivoting clamp frame. This way requires no welding or brazing because that's difficult for many people. The welded frame is my first choice because this method is a little less stiff and takes a little longer to make. However, they both get the job done equally well. Both styles will use the same top frames and both will interchange on the machine with no modifications to pivot location or brackets.

Let's start with a photo of a finished frame, then break down the build process. We start by making another aluminum clamp frame identical to the top one you already made. Then we attach a simple bent steel strip around it and extend it back to the pivot.

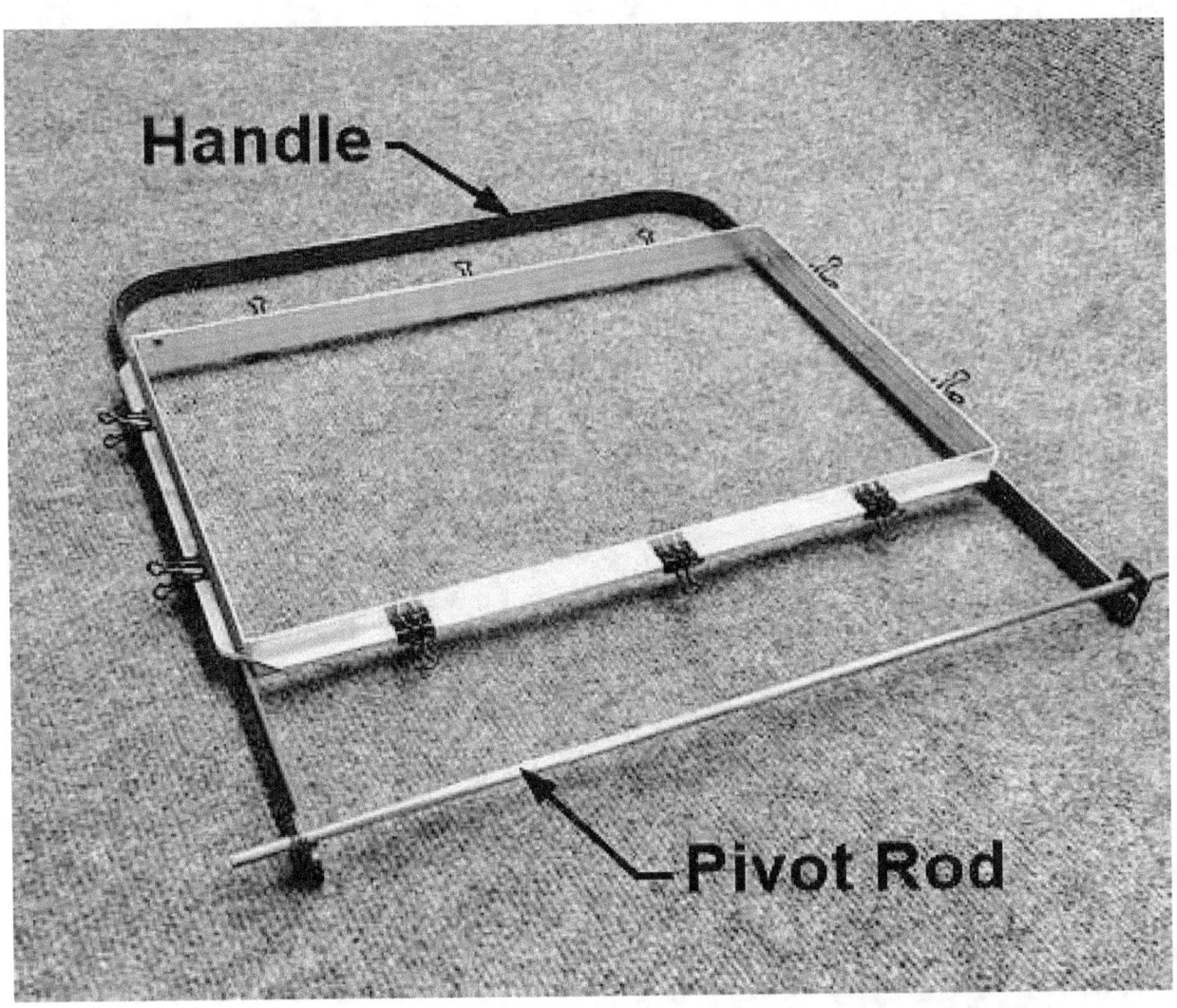

The steel strip is made from ⅛ x ¾ inch. flat strip. You can buy these at many hardware stores. ⅛ inch thickness can be bent by hand but requires some force.

It's hard to make bent parts with any precision so follow this sequence to make it easier. Make a wood form from 1x4 lumber cut to the dimensions shown and draw two 3 inch diameter circles. Trim the wood form and angle the end sides inward as shown to let you overbend. I chose the 13 ½ dimension which is undersized because it allows for springback when bending the metal.

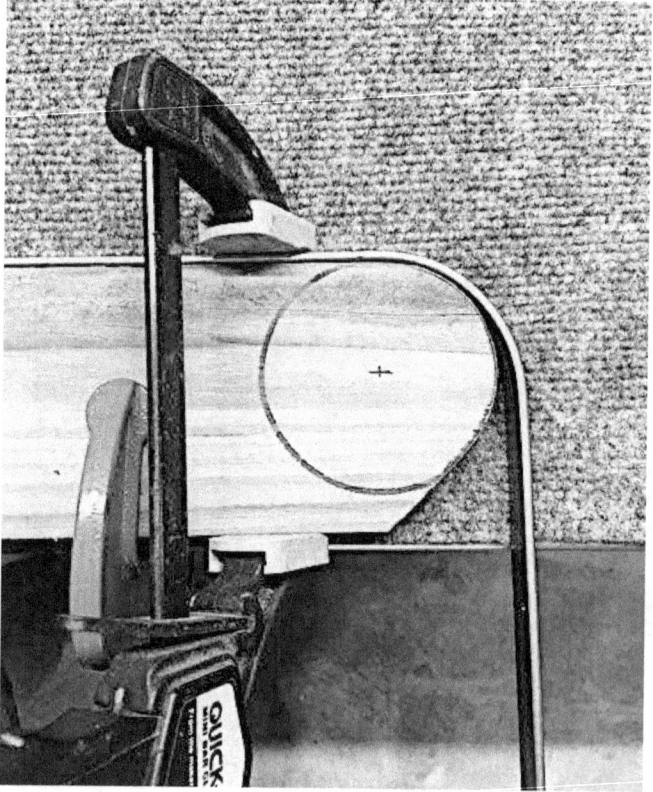

You can see on the left how much springback there is after bending and why the form is undersized and trimmed to allow you to overbend.

You can start by marking the center of your 6 ft steel strip and also the center of the wood form. Screw or clamp the form to a table or bench and clamp the steel strip to the form with the center marks lined up. Bend the strip around each end keeping pressure close to the bend and not pulling too much on the legs. You want them to remain straight. Over bend each leg until it springs back to a 90 degree angle from the front handle. Then massage as needed to get parallel legs with 17 ⅛ spacing between them.

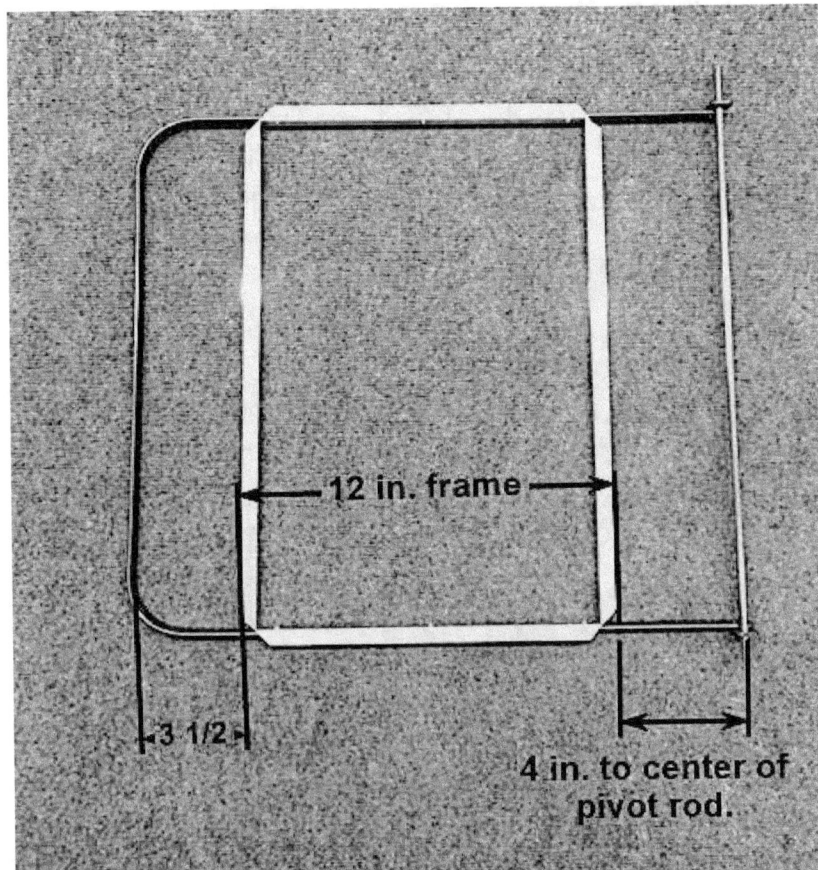

Place the clamp frame inside the bent steel handle. You can use the wood bending jig as a spacer so the handle is 3 ½ in. away as shown. Clamp them and drill three ⅛ inch holes for "Pop Rivets" or small 6-32 screws and nuts with the heads on the inside. The legs are left long until the tabs are located to put the pivot rod center 4 inches back from the aluminum frame.

The pivot rod location is critical so we will need to prepare two tabs as shown below with a ¼ inch hole and two smaller rivet holes. The 1/4 rod passes just above the main steel strip and is located 4 inches from the

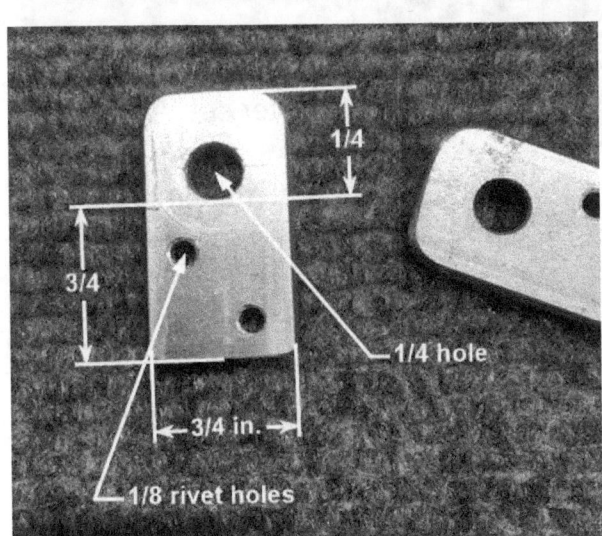

rear of the clamp frame to the center of the rod. When finished, the two legs fit between the wood pivot brackets. You can spread the legs if needed to keep it centered or use washers for centering. Use a ¼ ID shaft collars, or drill for cotter pins to keep the rod in place if too loose in the wood brackets

Chapter 6 - Sheet size Reducers

You can use plastic sheets smaller than 12x18 inches if you swap the platen for a smaller one and then build some sheet reducer frames. These will clip inside the larger frame to hold the smaller sheets. The logical sizes would be to use half sheets (9x12 inches) or quarter sheets (6x9 inches)

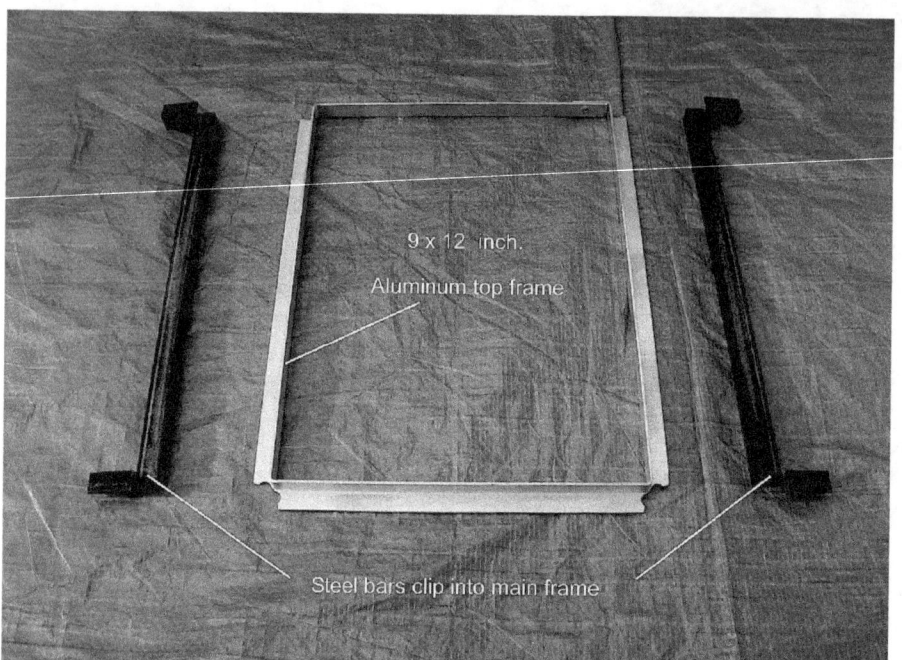

9 x 12 inch.

Aluminum top frame

Steel bars clip into main frame

This photo shows a 9 x 12 adapter set. The black cross bars are made of steel and clip into the main pivot frame,.space them 9 inches apart, then place the silver aluminum frame on top with spring clips.

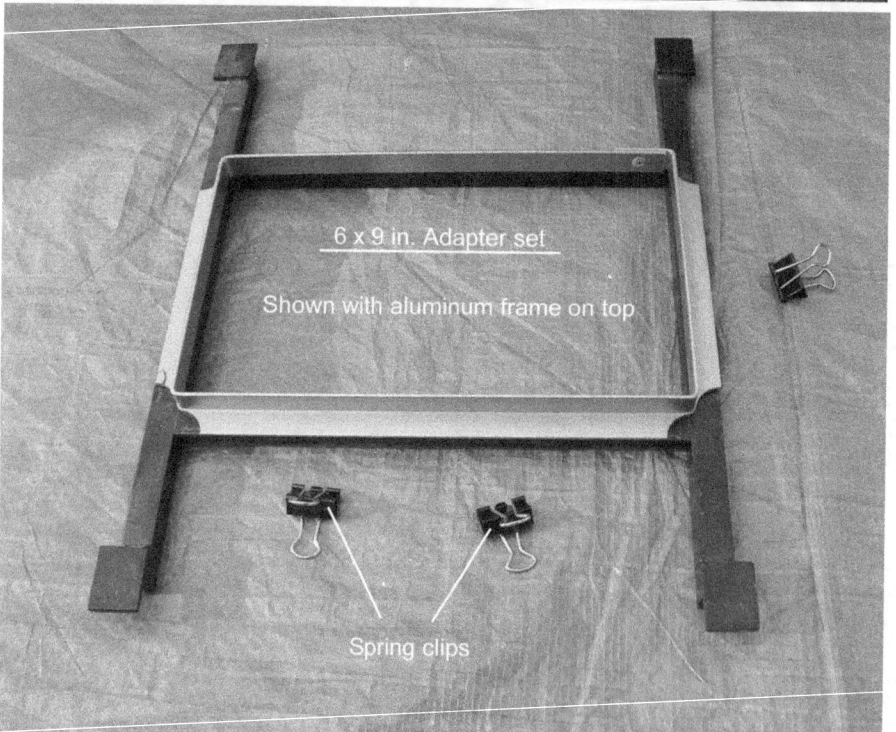

6 x 9 in. Adapter set

Shown with aluminum frame on top

Spring clips

This shows a 6 x 9 adapter set. Again, the black frame is steel and the aluminum frame (gold color this time) holds the plastic sheet flat with small spring clips as shown.
in bars as shown in the left photo above, or drawings # 12 and 13, then simply space them apart as needed for the smaller platen. They are held in place by a spring clip at each end.

This photo shows a 9x12 platen used with the two 12 inch adapter bars clipped in place. Just adjust everything to give a uniform gap around the platen and use the top frame that goes with it.

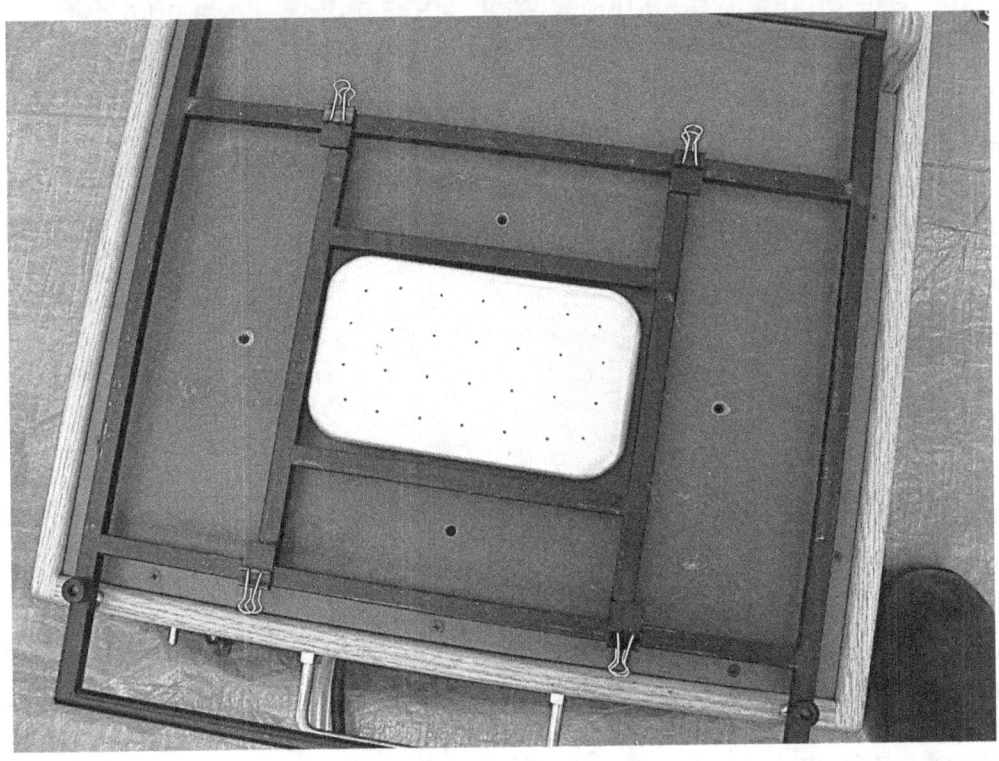

This photo shows the 6x9 inch platen and clamp frame adapter clipped into the larger frame.

Chapter 7 - Vacuum Pumps

This should be simple right? Melt some plastic and suck it over a mold. Actually, both heat and vacuum are very complex topics once you dig into them. Even at this primitive level of vacuum forming, the options for designing a vacuum system can get overwhelming. Making this more difficult is the challenge of keeping costs down and the changing market for vacuum pumps.

First understand that all vacuum cleaners or "Shop Vac's" are considered low vacuum pumps, but they also have a very high flow. In other words they are very fast but not very powerful.

When I first published these plans, real high vacuum pumps were very pricey so I came up with a two stage system to combine a vacuum cleaner with a very small high vacuum pump. I did this mostly because high vacuum pumps were too expensive so this was a way to make a very small pump work.

Things have changed!.... Imported vacuum pumps have dropped in price so much that you can now buy one large enough to have both good flow and high vacuum. Now you can build a much simpler "direct pump" system without the complication and cost of storage tanks or two stage valves. I chose to keep the info about two stage vacuum systems in this revision but to de-emphasize it. The simpler "direct pump" system with a larger (now cheaper) vacuum pump can perform as well with less complexity.

I know some people will find this topic confusing (or really boring) and others will appreciate the extra information. After all, vacuum is the primary force we are using here, we may as well try to understand it.

For those of you who are already getting a headache and don't want to become vacuum experts, read my recommendations below and then read only the section that applies to your choice. The rest of you can stick around and find out some interesting variations.

Pump Recommendations

Most popular Option- Buy an imported 4 CFM (cubic feet per minute) **or larger** rotary vane electric vacuum pump and connect it directly to your platen. This is called a "Direct Pump" system and it's the easiest system to build and use. The performance will be very good to excellent depending on your choice of pump (bigger is better). If this machine will be used by other people (schools etc..) then this would be the most user friendly solution. 90% of my customers use this simple system. No tanks or valves needed, just turn on the pump and flip the clamp frame.

Cheapest Option- A Shop Vac If you don't need much forming power and already own one. Otherwise put your money towards a real vacuum pump. Another cheap option is the hand pump/tank combinations with a dump valve. OK for occasional use and you can use it for oil changes too. Again, they aren't much cheaper than an imported vacuum pump.

Most interesting - A two stage system is technically hard to beat. Only a direct pump system with an oversized (7 CFM or larger) pump will match it. Both systems can finish off at full rated pump vacuum but the two stage system has a tiny edge in speed. This is where you combine a Shop-Vac with a rotary vane vacuum pump to boost the evacuation speed. Now with low cost larger pumps on the market it's not really worth the extra complexity and cost. I left it in these plans because someone may find it interesting and useful.

Vacuum - What is it? and how much do we need?

This may surprise you!! Vacuum or suction is really just the absence of atmospheric pressure. Most people don't realize this but the air all around us is pressurized to almost 15 psi. We are literally swimming in a sea of pressure and we don't feel it because our bodies are also pressurized. If you walked into a vacuum, you would explode because of your internal pressure (ouch!). When we vacuum form something, we aren't so much sucking the plastic down, we are actually removing the pressurized air underneath and allowing atmospheric pressure to push it down.

Vacuum is commonly rated in inches of water or inches of mercury. In either case vacuum is used to lift a column of fluid in a tube and more vacuum will lift it higher. Mercury is 13.6 times heavier than water, so obviously it won't lift as high. I will use inches of mercury from now on when I refer to vacuum levels, it's abbreviated IN. HG., let's just call it inches so I don't have to type it a million times. Keep in mind that if you see a pump rated in inches of water, (most vacuum cleaners do this) it's done for sales appeal because the higher number looks better. In fact, the only thing that uses inches of water are vacuum cleaners. It takes 13.6 inches of water lift to equal one inch of mercury.

Since vacuum is related to atmospheric pressure, we are limited to around 30 inches of mercury (equal to 14.7 psi) at sea level. You just can't get much more than that no matter how much you spend unless you figure out how to make the atmosphere heavier. In fact, as you go upward from sea level the air gets thinner or lighter so you can expect less vacuum (only 25 inches at 5000 ft.) It also changes a few inches either way with the weather or barometric pressure. Maximim vacuum is a moving target really and not a fixed value in our world.

How much vacuum do we really need?

The question of "how much" really has two parts. First we need to understand there is a difference between "Flow" and Vacuum level.. Using plain english,.. vacuum cleaners for instance don't suck very hard but they move or flow a lot of air. I would describe them as "low vacuum" but "high flow". They actually have more flow than we need and less vacuum than we want. It's a balancing act and you generally reduce one if you increase the other. Different types of pumps deliver vacuum in different ways.

In very general terms you want to reach full vacuum pretty fast, before the plastic chills. Generally again, you also want a high peak vacuum to suck the soft plastic down hard enough to capture details on your molds. Remember, the heated plastic is only soft till it cools so you need to hit it hard and fast with vacuum. More peak vacuum is desirable because it gives more forming power or detail in the finished part..

Some specific examples,.. For thin hobby and craft parts 5 inches of vacuum can get decent results. You might want 15 to 20 inches for better definition with more types and thicknesses of plastic, and 25 -29 inches to do more difficult parts or thicker plastics. Here are the most common types of vacuum pumps

Vacuum Pumps

Hand Pumps - Imagine a bicycle pump that works in reverse. You can actually take one apart, flip the piston over and install an external check valve to make a hand operated vacuum pump. The photos below show a similar pump you can buy for use as a marine bilge pump, and it's already set up for vacuum use.

Don't laugh, these can actually pull up to 28 inches of mercury. The obvious drawback is that they are slow and tiring. You can use one to evacuate a storage tank, but it may take 150 pumps.

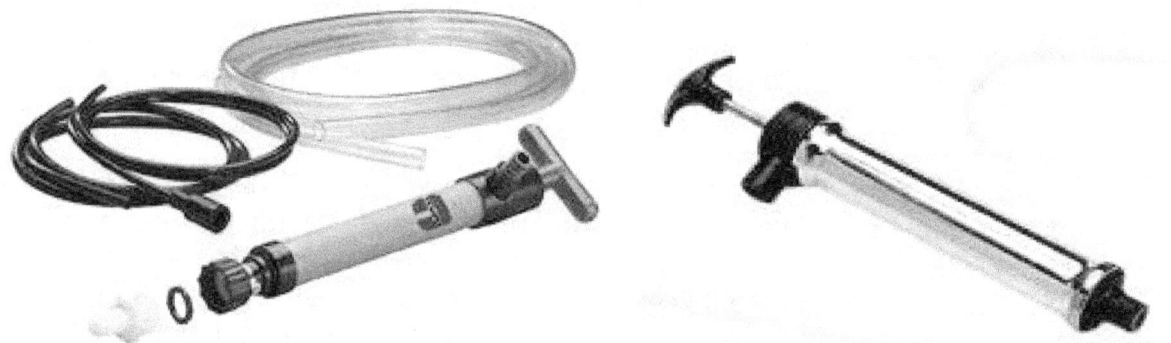

The two hand pumps shown above are marine bilge pumps. They are used for suction and come in several sizes.

In case you're wondering why not just use a bigger hand pump for more flow? The answer is that anything with a bore larger than about 1 ½ inches is simply too hard to pull. A hand pump will always need a storage tank because they are too slow. If you google" vacuum oil changers" you will turn up some interesting hand pump/tank combinations such as those shown below.

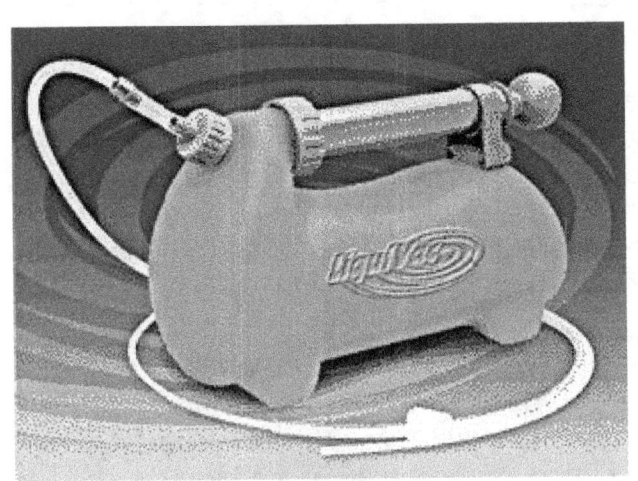

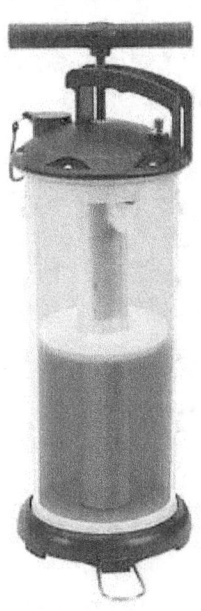

The manual pumps shown above are made for changing oil in a vehicle. You evacuate the tank and then suck the oil out. They seem perfectly suited to our use with the addition of a dump valve. You don't need much vacuum for changing oil but these can get up to 20-25 inches which is pretty good. These may be a quick way to get started with vacuum forming but before you get too excited, consider the limitations. These usually have short stroke pumps and small tanks so overall performance will suffer and they can be quite tiring to use. They can work well for what they cost.

Vacuum Cleaners - More accurately called Centrifugal pumps. These spin an impeller at very high speed and throw the air outward creating low pressure in the center. These move a lot of air fast but they don't pull very hard. You can only expect 4 to 6 inches from the typical vacuum cleaner. It's important to understand that even the biggest, noisiest 5 HP shop vacs won't do much better, they just flow more air. The very nature of the design limit's the vacuum level. Don't be fooled by 2 or 3 stage motors or high horsepower, or how much they dim the lights, They all pull about the same and more flow doesn't help us so a smaller shop-vac is fine.

Over the years many companies have tried selling low performance machines that use only vacuum cleaner motors and they still do. Sometimes the motor is built in so you can't see it but if it sounds like a vacuum cleaner don't expect much forming power no matter how much it costs. Except for light duty hobby and craft uses the results are almost always disappointing. These companies never publish the vacuum levels. The Hobby-Vac will obviously work with a shop vac but you can do much better.

Piston Air Compressor Pumps - These are positive displacement and simply pull air in one side and push it out the other. If you hook a hose to the output end you have an air compressor, and if you hook it to the intake side, you have a vacuum pump. That's right, all air compressors will also pull a vacuum. Some do much better than others because of little details in the design, but if you have an air compressor, see if you can tap into the intake port. A very few compressors are optimized for both pressure and vacuum and some even have a threaded intake port to screw a fitting into like the one below.

An air compressor can be used as a vacuum pump by tapping into the intake port

Don't worry about hurting an air compressor by using it this way, the piston doesn't know if it's pulling or pushing, it only feels like a 15 psi. pressure differential. Piston pumps are capable of pulling very hard up to about 28 inches, with some industrial pumps getting up to 29, but you should only expect 20 to 25 from your average converted air compressor.

Diaphragm Pumps - These are like the small air compressors sold for use with air brushes. Instead of a piston they use a flexible diaphragm. They are quiet and offer long life, but they don't perform too well for vacuum uses. You can expect only 10 to 20 inches of vacuum and most of them are too small for us to be concerned with.

Venturi Pumps - These simple devices allow you to blow compressed air through a venturi and create a vacuum. They are usually used in factories and can be tuned for high vacuum or high flow, but not both. It is possible to get multi stage unit's, but they are very expensive. They are also very inefficient, if your application calls for a 1 HP. vacuum pump then it would take a 5 HP. compressor blowing through a venturi to get the same performance. There is generally no good reason to go looking for one of these but 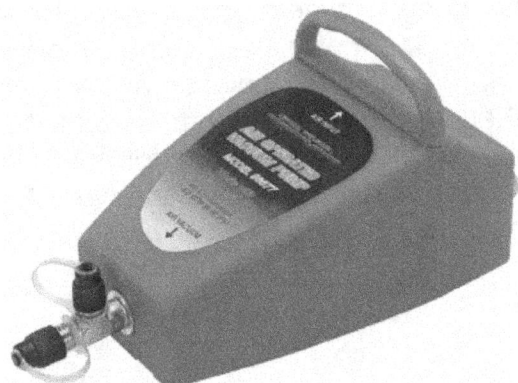 some imported ones for servicing air conditioning are getting cheaper. They are typically very low flow so need to be used with a storage tank or two stage system. Try one if you have one but there are better options.

.

Rotary Vane Pumps - The best choice,.. these use a spinning rotor with sliding vanes that rub inside a cylindrical housing. They are also positive displacement pumps and are capable of good vacuum and slightly better flow than a piston pump. This is how commercial vacuum pumps are constructed. They offer a long life and used to be very expensive. Some pumps have dry vanes that wear away very slowly and these will pull about 24 to 26 inches. Others are designed to be lubricated by oil and this type can pull 27 to 29+ inches of vacuum. Inexpensive imported Rotary vane pumps have flooded the US market and are the best value right now.

<u>Which pump is best for me?</u>

Over the years my customers have made it clear that they prefer simplicity and ease of construction. The overwhelming choice is to buy an electric vacuum pump big enough to run directly to the platen. No messing with tanks, hand pumps, check valves, plumbing etc,... it's the best all around for performance and value when you consider total costs and labor time. This can be a converted piston air compressor or better yet a rotary vane vacuum pump sold for A/C servicing. Here's what to look for in a Rotary vane vacuum pump

Min. 4-5 cubic feet per min. flow with ¾ Hp or more.

The pump to the right is a common type sold for evacuating air conditioning systems. They are often sold in 2 - 12 CFM sizes. Note that all Rotary vane pumps meant for servicing A/C systems will pull over 29 inches of vacuum because they need to. Just watch for the CFM rating and motor size. Some Chinese pumps exaggerate the ratings so avoid anything with less than a ¾ HP motor. Sometimes a 10 or even 12 CFM pump can be found for not much more money and these typically have a 1 HP motor and heavier weight.

The second choice may be a converted air compressor but these will not advertise vacuum ratings. They will also pull less max vacuum at around 22-27 IN.HG. The driving logic for these is that you may already have or want an air compressor. Look for a 1 ½ HP or larger piston type compressor with a threaded intake port. You may have to unscrew an air filter and install a fitting. The only way to determine max vacuum is to install a gauge and test it.

This info will help you while searching the web or on ebay. There is absolutely no problem with using a larger pump that flows more or pulls harder, these are just a minimum starting point for good performance with this size machine. Buy as much flow as you can afford. Sometimes the price isn't much higher and you may build a larger machine in the future.

If you are on a tight budget, already have a smaller vacuum pump or found a real bargain but it doesn't meet the requirements above, no problem. These plans will show you other systems that can make the best of it. The vacuum system is separate and upgradeable any time you want. Some people just use a Shop-Vac to start and upgrade when needed.

Pump Sources

Ebay, Amazon and Google

Air compressors can be found everywhere but you might be surprised to know that every day there are hundreds of rotary vane vacuum pumps for sale on ebay. Make sure to mis-spell vacuum every possible way to find those that were listed incorrectly. The challenge here is that many sellers don't supply enough information regarding the performance. All you can do is find the model number and do a web search to find out the specs. You can find new, used, surplus, and complete junk. It's a bit of a treasure hunt so you can win or lose big here.

Tip:
-- There are many ways to rate vacuum and flow, I prefer CFM and in.hg. but those are by no means all you will see when shopping. I suggest you Google the term "vacuum conversion chart" This will help you when searching for a pump.

Caution about power management!

Remember your Hobby-Vac machine heating element draws 13 – 14 amps depending on your exact voltage. This means it should be the only load running on a typical residential house circuit which is rated at 15 amps. Your vacuum pump should be plugged into a different circuit which may be in a different room. Furthermore, If you use a two stage system make sure your pump and vacuum cleaner combined don't exceed 15 Amps if they are plugged into one circuit.

Take the time to check the ratings on everything so you don't overload a circuit. Also avoid extension cords for the heating element. They can reduce power enough to cause problems. All pumps will be marked with their current draw in amps. It's also not hard to plug a lamp into an outlet then turn off a circuit breaker to see which outlets are on which circuits. As always its your responsibility not to overload a circuit.

Chapter 8 - Vacuum Systems

Like the previous chapter, this one has more info than you may need but a lot of people want that sometimes. Because the theme here is to keep it simple and this machine is a small one, I'll make the recommendation up front to use the "Direct Pump" system with a 4 CFM or larger rotary vane pump. 90% of the builders do this and I'll describe it first so you can move on if you're in a hurry. Those that have vacuum forming experience or have googled around a lot, will wonder why because the "pump and Tank" system seems to be the mainstream way. I will explain it all if you keep on reading.

Direct Pump Vacuum System

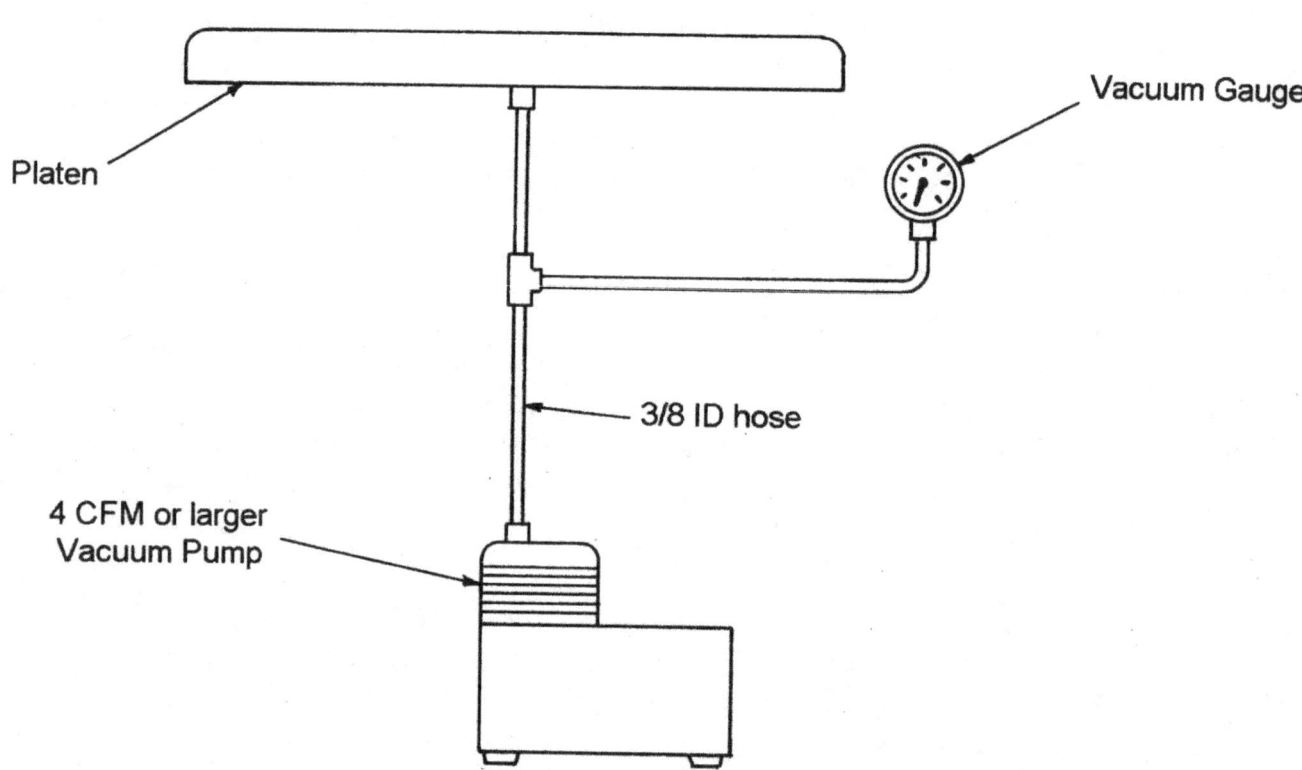

I suggested this system for several reasons. This smaller size machine works very well if the pump is sized right. It eliminates a lot of parts, cost and build time and makes operation simpler. And, over the last 5 years imported vacuum pumps have become plentiful and cheap. The 12 x 18 sheet size was chosen because it's the largest size we can heat effectively off a residential house circuit. With this size it turns out that a pump rated at 4 to 5 CFM or higher can actually be used by itself with no tank or dump valve. It's hard to get any more simple than that. You can take the money you saved by not buying tanks, dump valves, check valves or shop-vacs, and spend it on a larger pump. A larger pump is always desirable because it increases the flow and speed the plastic will suck down.

Use cost as your guide to purchasing a pump. They commonly make pumps in 4,5,6,8,10 and even 12 CFM sizes that still run on 120 volts. You may be surprised to find that a larger one may not cost much more than a smaller one.

Let me address the fact that some people may be hesitant to buy an imported pump. I'll stay out of the "buy american" debate and let you vote with your wallet. A 5CFM US made pump may sell for $500-$700 while a 10 or even 12 CFM imported pump may cost less than $200. Is there a difference in quality,... of course but not in proportion to the price difference.. If you were an air conditioning repair company, the US pump would likely last longer with better support for service and parts. However, that application demands that these pumps run for long periods to fully evacuate an A/C system. For our use, these pumps only run in short spurts of less than a minute so you don't deal with the high heat, wear and constant oil changes. What we do is really light duty for these pumps and cheaper import pumps work fine.

Advantages: The direct pump system is very simple to construct and easy to use, just turn on the pump and flip the clamp frame over. This is why it's so popular. A larger (faster) pump will make this system hard to beat. This system is also the most portable with just one external pump and no tanks or vacuum cleaners.

Disadvantages: Not many, you won't get the sudden explosive surge of vacuum like you can with a tank and dump valve. The plastic will form a little slower depending on your pump size, but this is offset by the fact that it finishes off at a higher vacuum level. Look at the performance chart and you can clearly see the difference between this and a tank system. Shallow molds are less sensitive to speed because there is less air under the plastic to remove.

To describe the performance of a "Direct Pump" system more specifically,,.. even a 2 - 3 CFM pump will work with some loss of detail due to unwanted plastic cooling. A 4 - 5 CFM pump is faster and will almost always work better than a tank system. A 7 CFM or larger pump will just about match the performance of a two stage system.

Read on if you are curious about other vacuum systems you may have heard about and see how they compare.

Helpful Tips:

-   *As the pump size increases, the vacuum line should also get bigger to avoid restricting flow. For up to a 5 CFM pump, use 3/8 dia. hoses from the pump to the platen. Increase to 1/2 dia. hose for 10 CFM etc,..*

-   *Always keep the hoses as short as practical to reduce internal volume and maintain forming speed.*

-   *Even small leaks in the platen or plumbing will reduce the maximum vacuum level a lot, so take this very seriously. All pumps drop to zero flow as they reach their max vacuum. This means even a pinhole leak can cause a major drop in your maximum vacuum level.*

Alternative Vacuum systems Explained

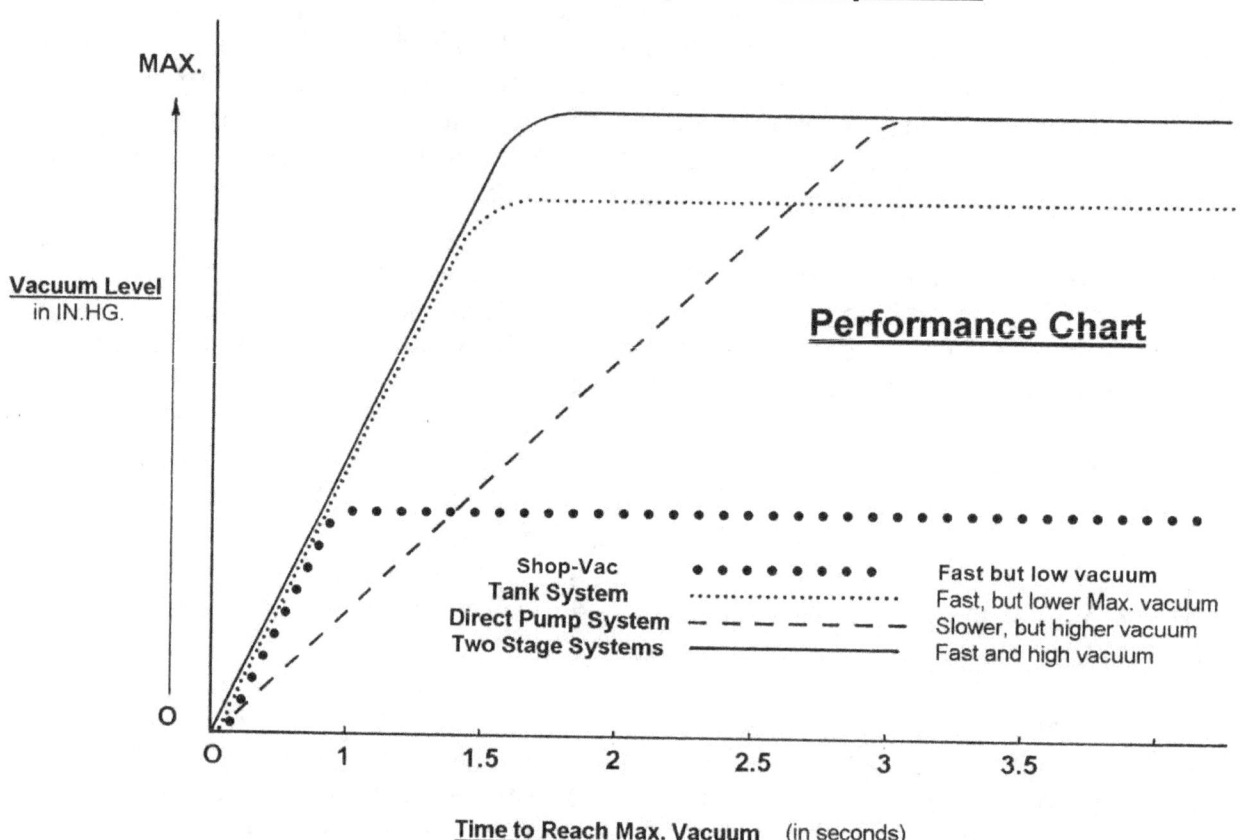

Above is a chart comparing speed and flow for four different systems. The numbers across the bottom represent time in seconds to reach full vacuum. On the left is the vacuum level with higher being better. Keep in mind that you only have a few seconds to form the plastic before it cools. Vacuum forming actually requires an odd combination of high flow first and then high vacuum.

Page - 54

In other words we want the plastic to pull down fast to do the heavy stretching while the plastic is softest, then we want it to pull down hard for final definition. As you can see, a **Shop-Vac** would rise quickly then level off at a low vacuum, its maximum forming power is only around 1/5th of the other systems being compared. I don't dwell much on the use of vacuum cleaners for this reason. It's obvious you can use a Shop-Vac on this machine but you can do a lot better with not much cost or effort.

With the **Tank System** you can see that the vacuum level rises rapidly as the dump valve is opened, but the maximum vacuum is lower than it could be. The wide dashed line is a **Direct Pump** system and you can see it finishes off at full rated pump vacuum but takes a little longer to get there. The speed is improved as you go with larger pumps with more flow.

The **two stage system** is both fast and powerful but is more complex and perhaps more beneficial in larger machines. The direct pump system is as powerful and only a little slower but with a much simpler system.

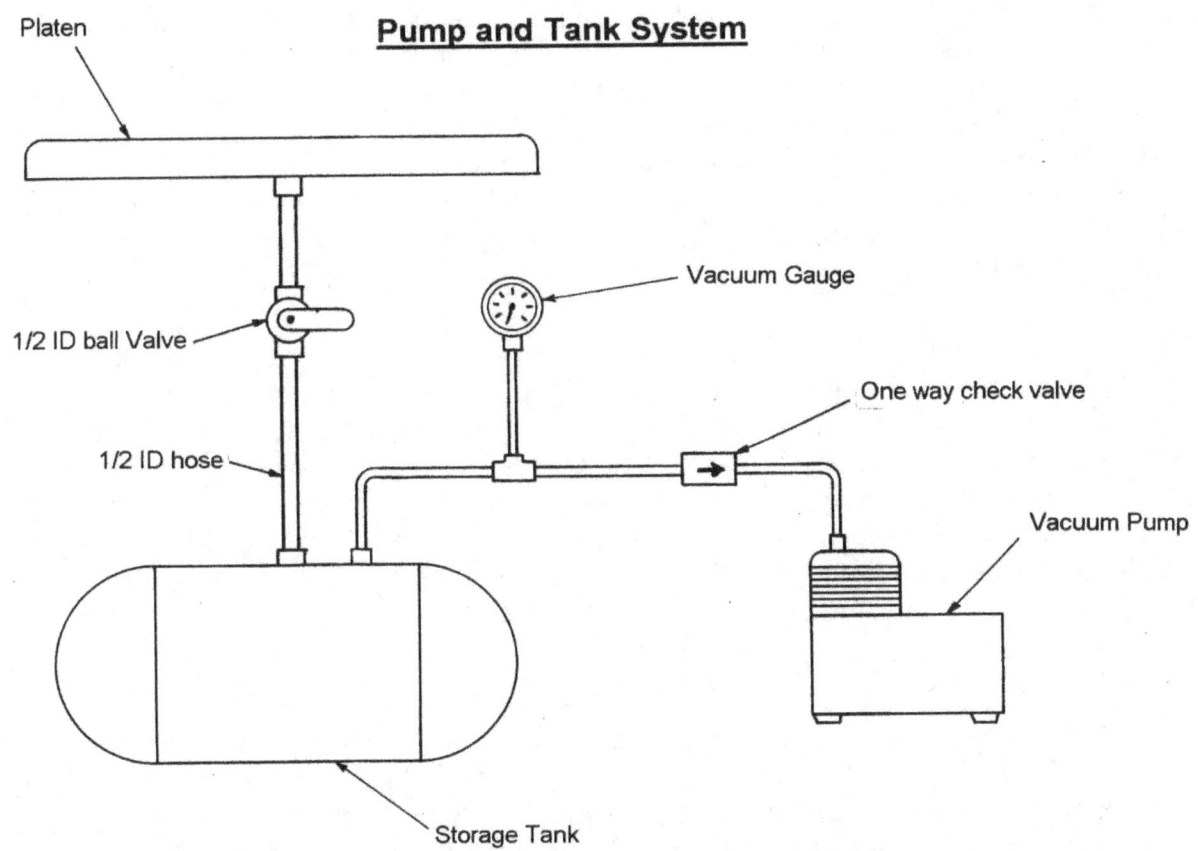

Pump and Tank System

Platen

Vacuum Gauge

1/2 ID ball Valve

One way check valve

1/2 ID hose

Vacuum Pump

Storage Tank

The pump and tank system is sort of the industry standard and only requires a short explanation. You simply close the main valve and evacuate the tank with your pump. When you are ready to form, just open the valve to "dump" the vacuum through your platen. This allows us to use a smaller pump to evacuate the tank slowly and then dump it quickly. This gives us the high flow and high vacuum we want. It sounds and looks simple, but it's really a series of compromises that ultimately causes performance loss. Virtually all commercial machines sold use this system and no one seems to care if it could be better. In fact, if I didn't tell you there was a better way, you would most likely be thrilled with the results. I'll explain the shortcomings and make some suggestions for building a system like this.

Advantages: Easy to understand and the ability to use smaller (cheaper) pumps than a "Direct Pump" system because it gets its speed from a stored vacuum and dump valve. This system is very tolerant of leaky platens and mis-matched components. Just throw some parts together and it will always work OK. This probably explains it's popularity more than anything

Disadvantages: In a standard pump and tank system, the vacuum level that you finish with is always lower than the pump's rated vacuum. For example, let's say your vacuum pump can pull 29 inches. When evacuating a tank the vacuum level drops quickly at first and then gets real slow as it reaches maximum vacuum. It may take two minutes to get to 23 in. and another two minutes to get to 25 in. By then it's so slow you just give up the last inch or two. As the part is formed, air from under the plastic sheet rushes into the tank and you lose even more stored vacuum. By the time you are done, you may have only 20 or 25 inches left in the tank even though your pump is capable of 29. It doesn't help to leave the pump running because the tank slows it down too much. Despite these shortcomings, tank systems get the job done and tolerate leaky systems. The performance can be improved a little by throwing more money at it and buying larger pumps and tanks.

Tanks: This machine is small so the components are relatively cheap. I would go with a tank capacity of 5 to 10 gallons. Even a 2 CFM pump will evacuate a 5 gallon tank down to 25 - 26 inches in about 2 minutes, but you should expect at least 2 inches of vacuum loss between parts. A 10 gallon tank with a 4 CFM electric pump will pump down to 25 - 26 inches in about the same time and will lose around one inch of vacuum when a part is formed. Note: You must use a one way check valve between your pump and tank so you can turn the pump off and still hold vacuum. You can buy small 5 to 10 gallon air tanks used for filling tires at any auto parts or hardware store.

Caution!
Any tank must be sturdy, such as a portable air tank for filling tires, an old propane or Freon tank (empty thoroughly), or a small air compressor tank. Do not use glass containers that can break or plastic jugs that can collapse. NEVER anything with residual flammable or corrosive liquids or gasses inside.

Dump Valves: The main dump valve should be something that opens quickly and flows freely. A quarter turn ball valve fit's that description and can be mounted to the front of the machine as shown below

Ball Valve Location
On Front Panel

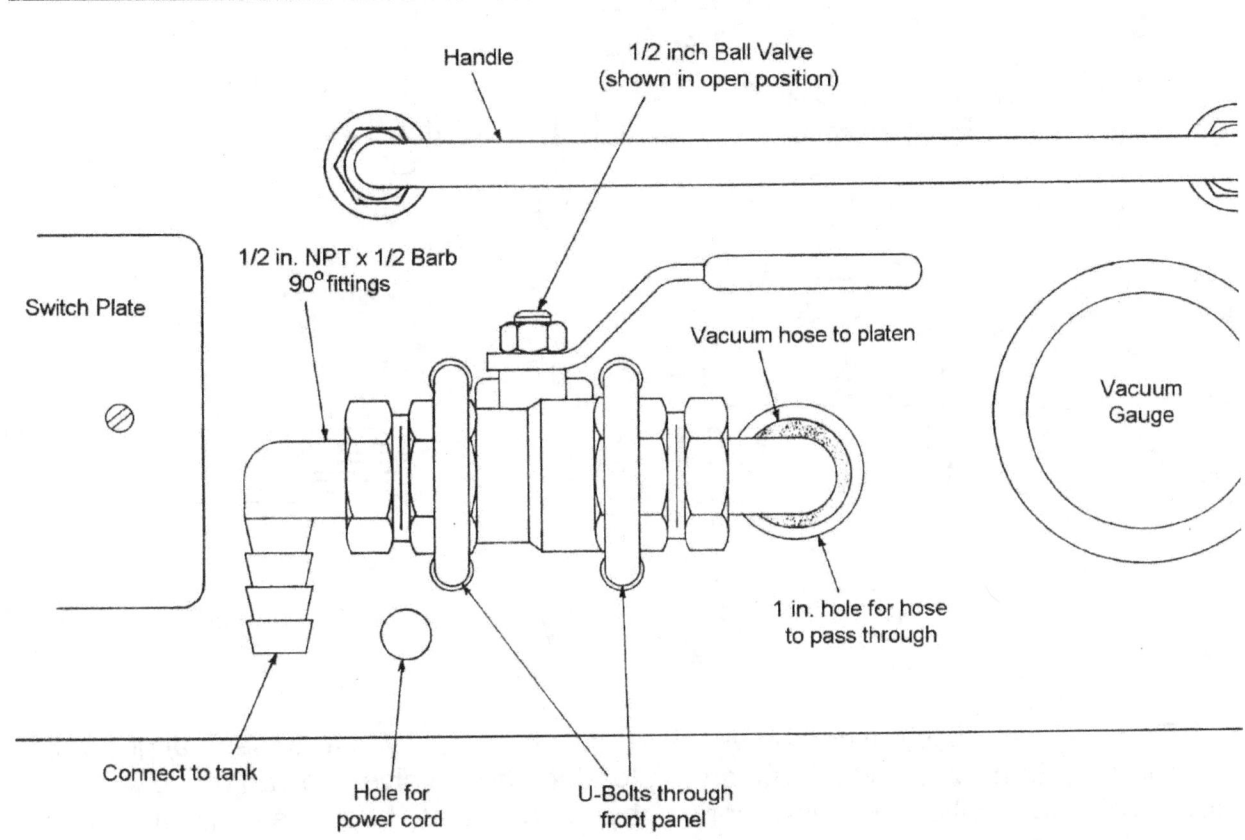

Handle

1/2 inch Ball Valve
(shown in open position)

1/2 in. NPT x 1/2 Barb
90° fittings

Switch Plate

Vacuum hose to platen

Vacuum
Gauge

1 in. hole for hose
to pass through

Connect to tank

Hole for
power cord

U-Bolts through
front panel

   A suitable 1/2 in. ball valve is shown in the parts sources. Use the optional lower handle so you can hold pressure on the clamp frame and open the valve with your free hand.
   Other valve types won't work as well, A ball valve has a smooth bore when opened for low restriction to flow. If you buy one at a store, try them all to find the one that opens easiest. These are typically used for gas shutoff valves and are easy to find. Use 1/2 ID fittings and hoses, and a 1/2 in barb fitting on your platens.

Automatic Two Stage Vacuum System

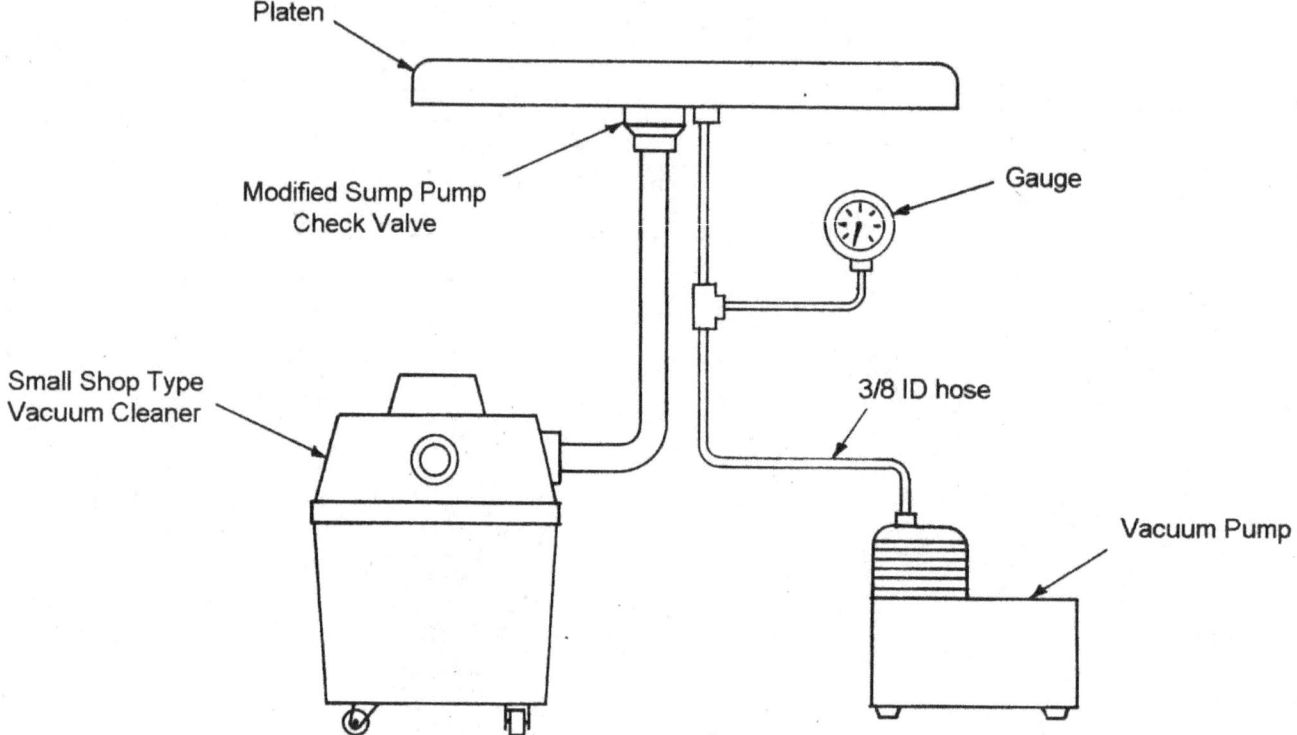

The two stage system combines the speed of a Shop-Vac and the final pull of a direct pump system. By using a large spring loaded check valve, we can make the system totally automatic. The performance chart shows what happens (solid line). It does add to the building time and it must be done carefully because it is very sensitive to leaks.

Do I really need to go to this much trouble?

This clever concept was more relevant when I wrote the plans many years ago. At that time even small vacuum pumps were expensive so it was worth the time to use the smallest one possible. Now it's not much of a benefit for the extra complexity. Not everyone needs this level of performance. In fact, the added forming power doesn't really show up until you start doing difficult pieces such as thicker plastics and deep draws. I am leaving this info in the plans as a bonus and for those who are curious I'll offer a detailed explanation.

Two Stage Operation

The heart of the system is a simple one way check valve with an ordinary shop type vacuum cleaner attached to it. This becomes your first stage and provides the high flow or forming speed we need. Shop-Vac's can flow over 100 CFM so if we stopped here, you would have a system that was crazy fast but only pulled 4 to 6 inches .

Your second high vacuum pump can be anything but in this case is a small rotary vane pump capable of 29 in.hg. It is connected directly to the platen and will provide the final power we need. With both pumps running, the operation is fully automatic. The check valve senses the flow to the vacuum cleaner and when it can't pull any more, a light spring pulls the valve closed. This allows the second stage pump to take over and pull harder against the last little bit of air remaining. It does this without being slowed down by a tank. Since most of the air was removed already, your second stage pump can reach its maximum vacuum quickly. The whole process happens in less than one second and you can't even see the transition.

Tip: Your platen must be air tight for the second stage to be effective. This means no wood in contact with vacuum. No matter how well you seal it, wood still leaks. You will notice this design uses no porous materials in its construction.

Selecting a Check Valve

The key to operation is the spring loaded one way check valve that senses first stage flow and closes when the Shop-Vac reaches it's maximum.

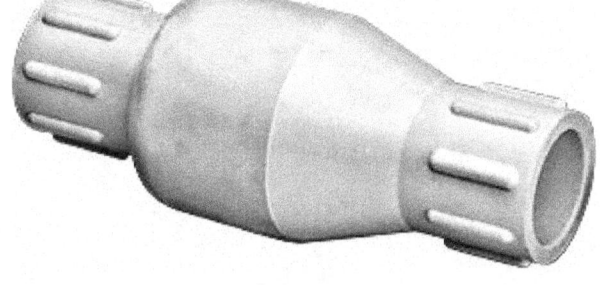

The valve we need must have low restriction to flow for the first stage, and the valve must be held closed by a very light spring. Fortunately, you can buy something that does just this and it's used as a basement sump pump check valve. Valves of this type can be found in many hardware or builders supply stores but very few have spring returns like this one.

McMaster Carr # 46835K54 with 2 lb return spring

Others may work as well, but the important thing is that the valve must close well and form an airtight seal. The spring must also be very light. You can even check these things in the store by blowing into one end and then the other. It should open easily in one direction and close perfectly in the other. Not all valves will seal perfectly. The other basic requirement is that the overall length with a 90 degree elbow installed must be short enough to fit inside the cabinet. Or else you can raise the machine up with feet to make room.

Modifying the Valve

The check valve wants to be mounted so the actual valve is very close to the platen to reduce internal air volume. The drawing below shows how excess plastic can be trimmed away and glued back together to shorten the overall length to fit inside the cabinet better.

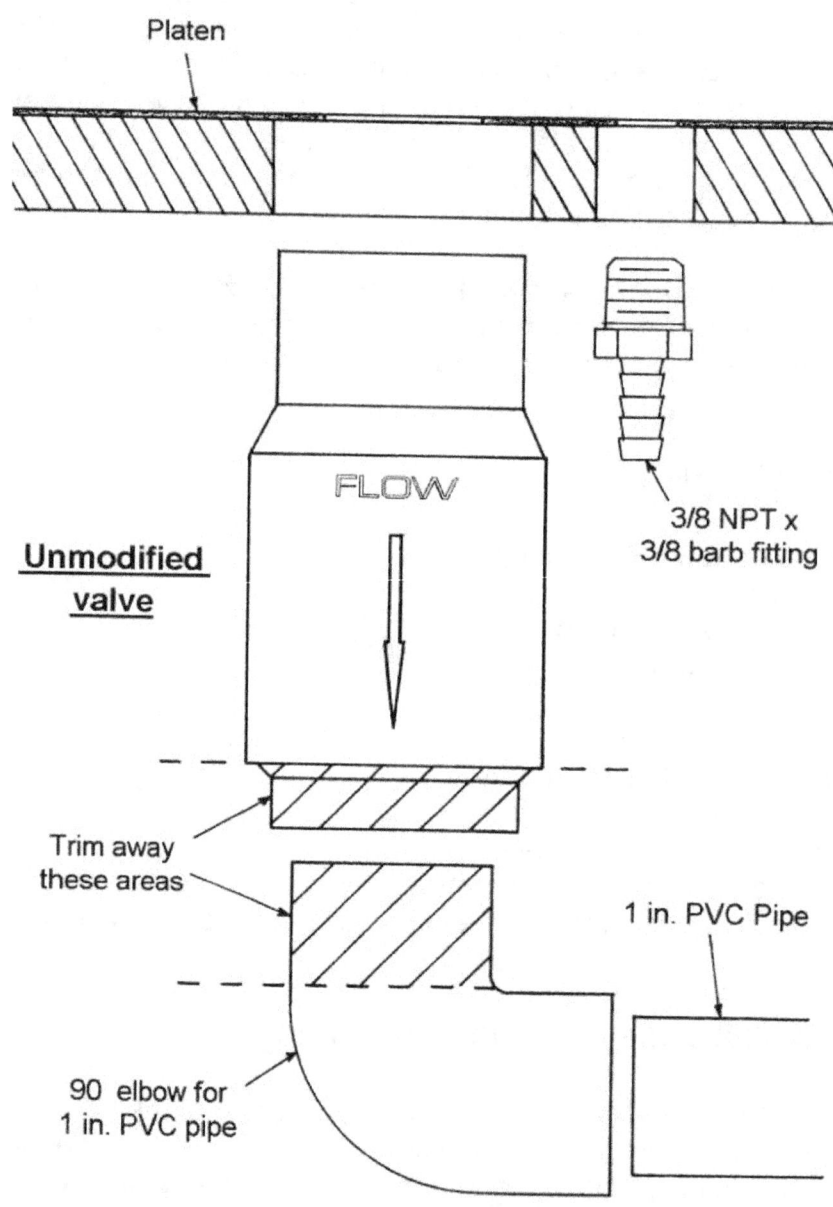

The drawing below shows the same valve shortened and installed inside the machine. Note you can also trim some off the top if you need to make it shorter. Also note that the valve is pushed up through the wood particle board platen and is glued to the bottom surface of the Formica laminate top layer. The same principle applies to the small fitting. This method provides no option for vacuum to leak through the porous wood platen

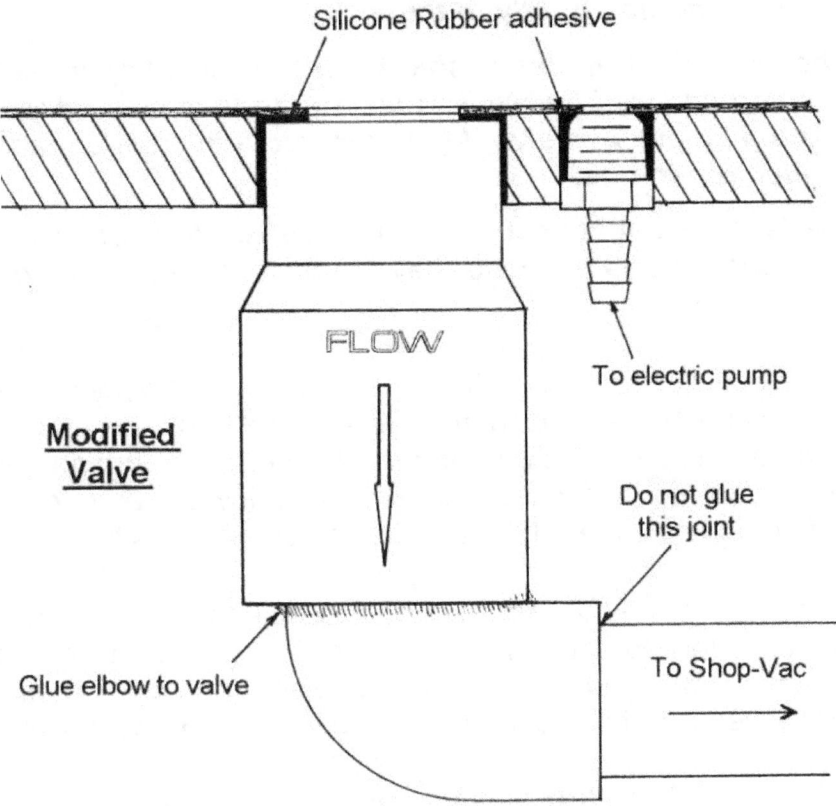

Silicone Rubber adhesive

FLOW

Modified Valve

To electric pump

Do not glue this joint

Glue elbow to valve

To Shop-Vac →

Modified Check Valve
For automatic two stage systems

Notes:
-- This valve is available from McMaster Car
see parts list.
-- This valve is spring loaded closed

A length of straight PVC tubing will extend through a hole in the front panel and connect to the hose on your vacuum cleaner. A short section of 1 1/4 ID radiator hose makes the perfect coupler, but a rigid PVC coupler will also work. You may have to wrap the end of your vacuum cleaner hose with tape so it fits snugly.

Tips:

-- Don't glue the straight pipe into the valve, it must be removable so you can change platens.

-- The smaller platens don't need a two stage valve, they have so little volume that you can run your pump directly to them.

-- Everything between the hot plastic sheet and the valve flap must be absolutely air tight. Even a pinhole sized leak will cause you to lose several inches of vacuum. Everything below the valve flap to the vacuum cleaner can tolerate leaks and not make much difference.

-- Even the smallest, cheapest vacuum cleaner will do this job well, don't spend more money for a bigger one because they are all faster than we need.

Advantages of a Two Stage System: A very versatile system, disconnect the vacuum cleaner and use it as a direct pump system, and hook up the shop vac only when you need the extra speed. No operator involvement, just turn on both pumps and flip the plastic over. Can effectively use smaller pumps down to around 2 CFM. such as a venturi pump. Some people already have small electric pumps for vacuum bagging.

Disadvantages: The extra cost and work to modify and install a valve, and you must pay extra attention to avoiding leaks. You must choose a pump and vacuum cleaner that won't exceed 15 amps combined and run them on a separate circuit from your heating element. You have the bulk of two external pumps to clutter up your workshop.

Summary of Vacuum systems

Use a Shop-Vac for thin simple parts that don't require much definition. Use a direct pump system for the most cost effective high performance system. Use a tank system as a way to use a smaller pump you may already have, or if you have a leaky platen. The two stage system can offer top performance and also works with smaller pumps if you already have one but it is very sensitive to a leaky platen.

The direct Pump system with a pump greater than 4 FM is the most popular.

Chapter 9 - Building the Platen

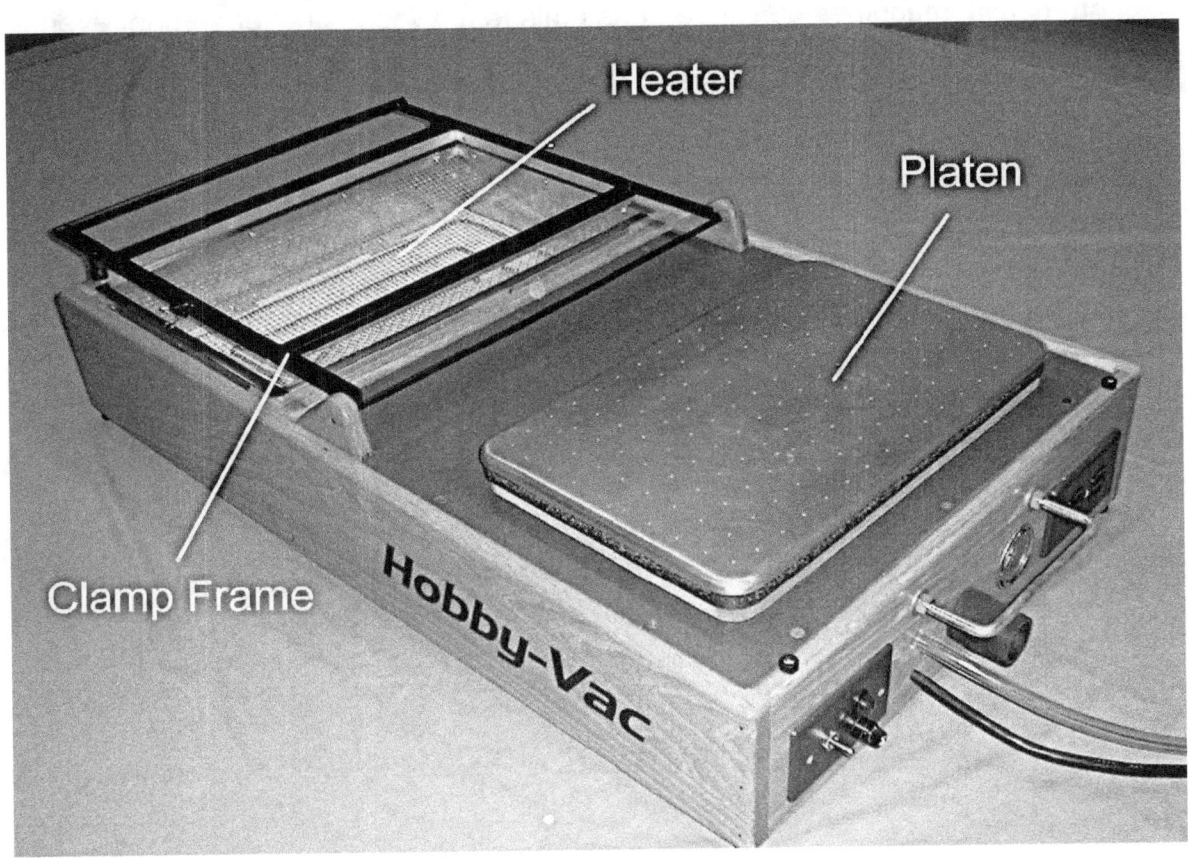

The "Platen" or forming surface is where you place your mold. I think the best type is one that has a perforated surface with tiny holes. This allows multiple molds to be placed anywhere for maximum versatility. It has three main purposes. It must support the weight of your mold or pattern. It must allow vacuum to reach all of its surface area, and it must provide a way for the plastic to seal around its perimeter.

Design Goals

This is where most home built machines fail miserably. There are a million ways to make a platen, but the worst way is to use a wood box with a pegboard top, yet this is how it's usually done. A wood box is never air tight, has way too much volume, and the holes are much too large. If your vacuum system has really good performance it can suck even thick plastic through a 3/16 hole until it pops. Since the platen is usually the weakest link on a homebuilt machine, I decided to go all out and incorporate the following desirable features into this platen design.

- **100% Airtight Construction** - Ideally, if you were to lay a rubber sheet on top of the platen and apply a vacuum, there should be absolutely no leaks. Most commercial machines can't even pass this test. Wood is never air tight even if sealed with paint so we should use non porous materials where needed.

- **Minimum Internal Volume** - This is the most common mistake. Any extra space inside the platen is a bad thing. It takes precious seconds to evacuate it while the plastic is cooling. This design has only a 1/16th. inch space between the layers and the volume is reduced even further by the screen inside. I consider this the minimum volume that can still provide good flow to the corners. I have never seen a commercial platen this optimized. The low volume means fast response to vacuum and this is essential when using a two stage vacuum system.

- **Multi-Size Capability** - Make as many different size platens as you need. They simply screw down to the top of your machine and can be changed in minutes with the same outstanding performance regardless of size.

Choose a Basic or Deluxe Platen

I'm going to show you two ways to make a platen and they both perform exactly the same. The Basic way is just cheaper and easier to build, but it requires occasional replacement of the perimeter gasket which will develop leaks if not maintained.

The Deluxe platen goes a step further and adds a perforated aluminum top sheet with smooth rounded corners for the plastic to seal to. The Deluxe version is maintenance free with no exposed seal maintenance and is the preferred method if the machine will see a lot of use. You might also use both types, for example, you can make a deluxe 12 x 18 platen that gets used most of the time and smaller basic platens that get used less frequently. I highly recommend that you make three platens so you can form full, half, or quarter sheets. The materials don't cost much and once you have the tools out, It's easy to make all three. You can even make the smaller ones first, just for practice

Basic Platen Example

Shown is the 6x9 platen. Simple and fast construction. The foam rubber seal must be butted tightly at the corners.
The screen is cut to fit inside the seal and it allows vacuum to travel freely under your mold to reach the full area even with multiple molds.

Deluxe Platen Example

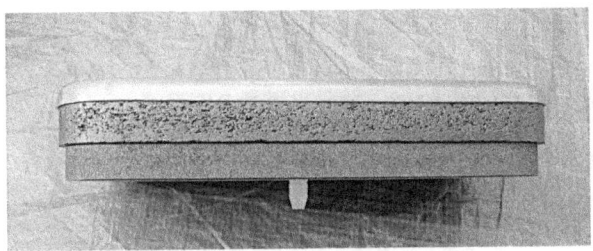

Shown above are the half and quarter sized deluxe platens. The wood base (painted brown) has rounded corners and the soft perforated aluminum top sheet is bent over with a soft mallet.
The screen is sealed inside between layers. The hot plastic sheet seals to the rounded corners with no rubber gasket needed. Note that the corners are also rounded when looking down from the top. Without the square corners it allows the platen to miss the plastic in the corner of the clamp frame that is stiffer from uneven heating. These details are not much more work and together they make it a better all around performer.

The basic and deluxe versions share the same materials and cut sizes except for the aluminum top sheet which we will discuss later. The following chart gives overall part dimensions for all three platens. Full 12x18,.. half 9x12,.. and quarter sheet 6x9. Please note that although the parts for both platen styles are the same rough dimensions, the "Deluxe platen" has all the corners trimmed with a radius as explained in that section. Cut the parts with square corners for the "Basic platens"

Platen Part Dimensions

Platen Size	Small 6 x 9	Med 9 x 12	Large 12 x 18
Bottom Spacer	4 ¼ x 7 ¼	7 ¼ x 10 ¼	10 ¼ x 16 ¼
Platen Base	4 ½ x 7 ½	7 ½ x 10 ½	10 ½ x 16 ½
Screen	3 ½ x 6 ½	6 ½ x 9 ½	9 ½ x 15 ½
Aluminum top sheet (for deluxe platen)	5 x 8	8 x 11	11 x 17

Bottom Spacer - This just raises the platen up to the correct height. It can be any 1/2 inch thick wood. The spacer board is a little smaller than the platen itself and can be fixed to the machine with screws up from the bottom. It can be permanently screwed or glued to the underside of the platen. The center hole size is not critical, It just needs to be large enough to clear whatever fitting or two stage valve you are using.

Platen base - This is 3/4 in. thick particle board with a top layer that air can't pass through. Believe it or not, a strong vacuum will pull air right through any type of wood, even paint or varnish won't form an effective barrier. To create an airtight barrier layer, we can use a thin sheet of aluminum or Formica. A steel sheet can rust and plastic can't deal with heat exposure. You can cut up an aluminum cookie sheet or pan and glue it to the wood with contact cement, construction adhesive, silicone or urethane adhesive.

Tip:
Another easy choice if you can find it is to use a piece of kitchen countertop material. This is actually a dense phenolic substance about 1/16th. thick with a decorative top layer. This thin laminate is glued to the particle board for support .

A common trade name for this product is "Formica", but there are others also. If you can find a company that installs kitchen countertops, (look in the phone book) then you can ask for their sink cut-outs. You can usually get all you want for free or at least buy some scraps from them. Color and pattern are unimportant, but make sure the laminate top layer is around 1/16th. in. thick. Some countertops use a thinner 1/32 in. thick laminate that is less durable. You should also avoid a material commonly referred to as "Melamine" which is thinner yet and is typically used on shelf boards and cheap furniture.

Screen - This is not critical and it only serves to keep the plastic from sealing off around a mold on a Basic platen so vacuum can travel to the whole surface. A Deluxe platen has screen under the top layer. Window screen is too thin and hard to keep flat. The best screen is often called "hardware" cloth" and is sold in hardware and lumber stores. ¼ inch mesh is best but ½ inch mesh can also be used.

Holes for small platen fittings

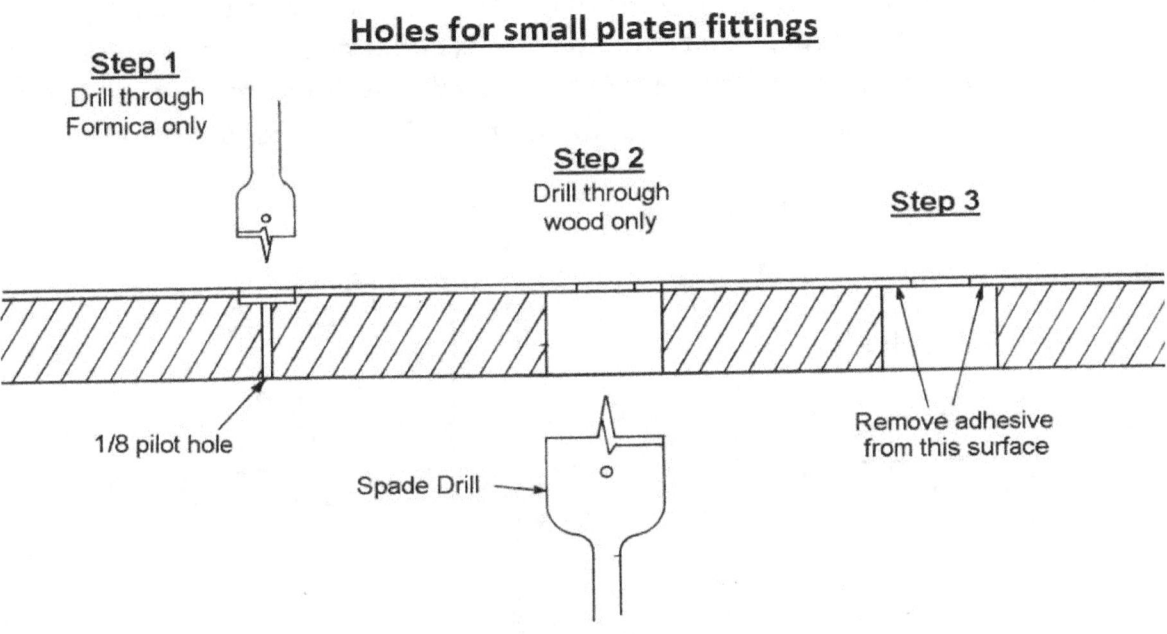

Step 1
Drill through
Formica only

Step 2
Drill through
wood only

Step 3

1/8 pilot hole

Spade Drill ➝

Remove adhesive
from this surface

Vacuum Fittings - The small fittings should be 3/8 NPT male thread on one end and 3/8 or 1/2 in. barb on the other end, depending on what size pump you are using (read chapter 7 first). The threaded end of the fitting will fit loosely into a 1 inch hole, with a 1/2 inch hole through the laminate. Choose the appropriate hole saw sizes if you will be using a two stage check valve. Remember, you only need the check valve for the Large platen if you choose to go that route. The two smaller platens won't benefit from two stages and can always use a single smaller fitting. For small fittings, the holes can be made with two sizes of spade drills.

For larger fittings or check valve installations, the concept is the same except the holes are larger and can be made with hole saws. The end result is sort of a receptacle in the bottom of the platen for the fitting or valve to glue into. You can use plastic or metal fittings. Several types of adhesives can work to secure the fittings. 100% silicone rubber caulk, newer urethane adhesives and epoxy putty adhesives. Avoid latex caulks and any construction adhesives or glues that get brittle.. Make sure you clean the back side of the top sheet so the adhesive seals well there. The glue along the sides of the fitting is just there for support, but it's absolutely critical that there be no leaks between the fitting and the top layer of aluminum.

Important

Whether it's a small fitting or a larger check valve, the installation method is similar. You will notice from these sketches that we want a large hole through the particle board and a smaller hole through the aluminum or Formica top sheet. The important thing to know is that the fitting must glue and seal to the top sheet to retain vacuum and prevent air from traveling through the porous wood base. Gluing it to the wood sides is just there for support. The critical seal is at the top sheet!

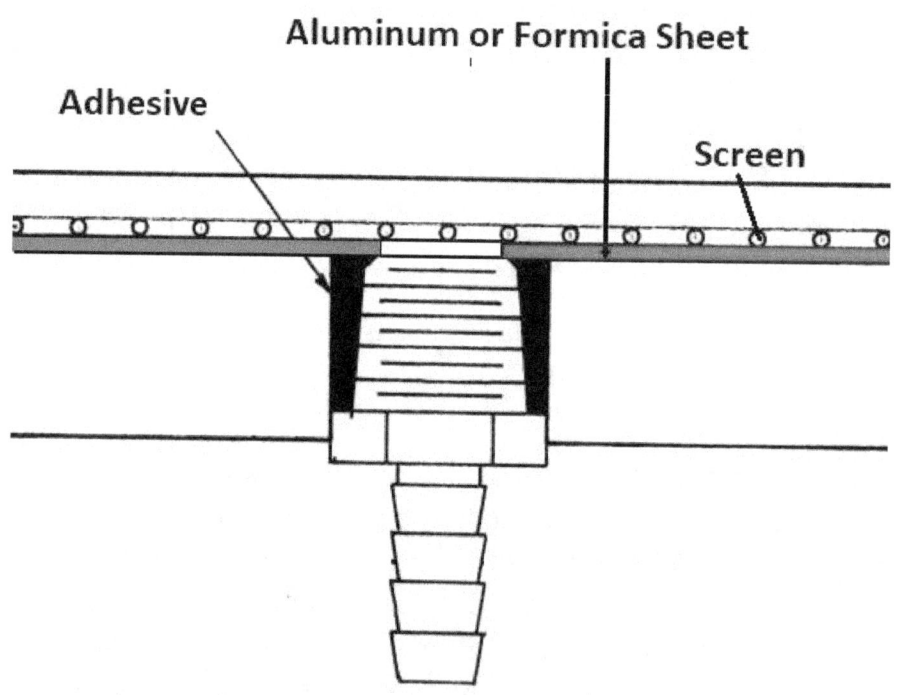

Building the Basic Platen

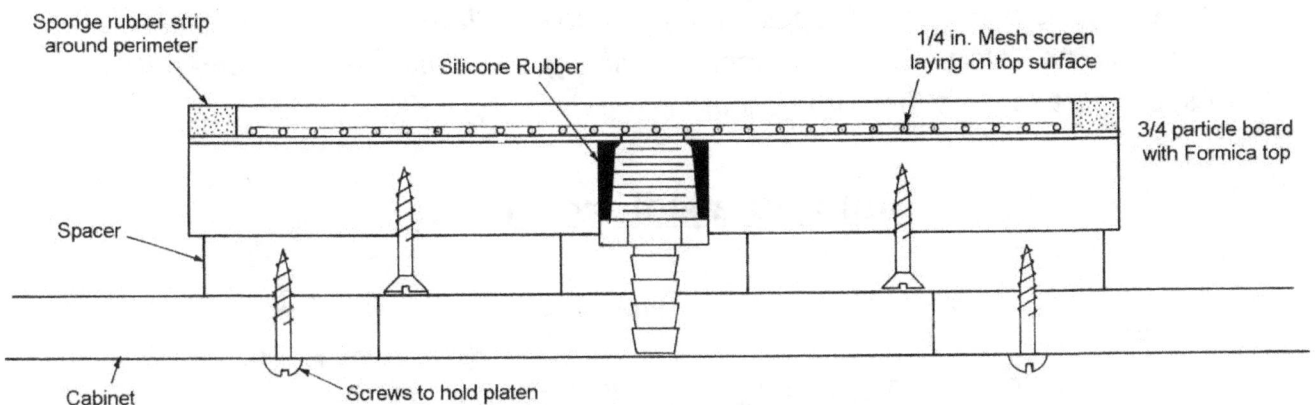

All of the materials mentioned so far are the same for both types of platens, let's assemble a basic one first. If using a "Direct Pump" system, all of your platens will use a single small fitting. If using a Shop-Vac or "Two Stage" system your large platen will use a larger 1 inch PVC coupler for the Shop-Vac, or a check valve will be installed.

Cut all the parts to the sizes shown on the chart and glue your fitting into the platen base and let it cure overnight. Screw the spacer board to the bottom of the platen, but make sure the screws don't poke through the aluminum top. Apply a strip of self adhesive sponge rubber around the perimeter then lay the screen in the center and you are ready to go. Fasten the platen to the machine with screws up from the bottom or optional threaded knobs

Selecting a Foam Seal -

There are many types of self adhesive foam rubber strips used for weatherstripping available from your local hardware store, but here's what you should look for. Use a firm grade of closed cell rubber foam, (open cell foam will let air pass through). Try any closed cell foam you can find, but generally something made of Neoprene, EPDM, SBR or Silicone will work well. Try to avoid PVC, Vinyl or Polyethylene foams because they don't hold up as well but can still work for a while.. The size should be 1/8th. to 1/4 in. thick and 1/4 to 3/8 in. wide. Peel off the paper backing and stick the foam strip to the aluminum or Formica top as shown. Make sure the corners butt tightly together.

- A light coat of petroleum jelly (Vaseline) on the foam rubber strip will keep the plastic from sticking and should make it last longer. You may have to try a few different types of weatherstripping until you find one that holds up well. Clean the aluminum or formica so the foam adhesive grabs better. You can use a drop of super glue to keep the corners joined.

Building a Deluxe Platen

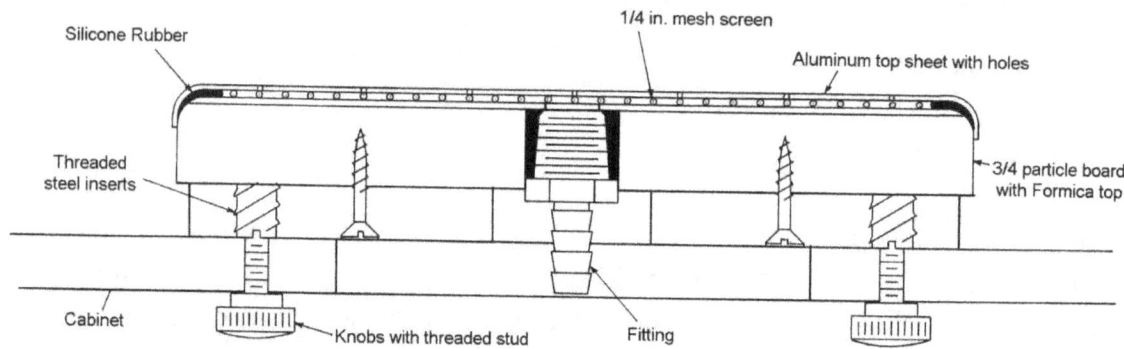

The Deluxe platen has some nice refinements. First there is the lack of a perishable foam perimeter seal. Instead we have a smooth round corner that seals perfectly against the soft hot plastic. Also looking down from the top, the Basic platen has square corners and the Deluxe version has a generous radius on all four corners. This is nicer because the plastic sheet always suffers from having colder corners than the center of the sheet. The Deluxe platen pushes into the hot plastic easier because it's not fighting the stiff corners. Another nice touch is the threaded knobs and inserts used to hold the platen in place. Lastly, the formed plastic outside the mold will have a flatter surface without having a screen texture. I think there are enough small improvements to warrant a little extra work over the basic platen.

Cut the wood pieces for each layer the same way as the Basic platen with the same dimensions as shown in the chart. You will also need to trim all the layers with the large corner radius as shown on the following page. You can then create the smaller ¼ inch radius on the entire top perimeter of the wood base. A router is best but this can be done by sanding or filing too.

The top sheet is a soft aluminum sheet trimmed to the specs on the chart and is pre-drilled with 1/16th holes on a one inch grid pattern. The platen base is wood with Formica or another aluminum glued to the top surface

Layers for a deluxe Platen

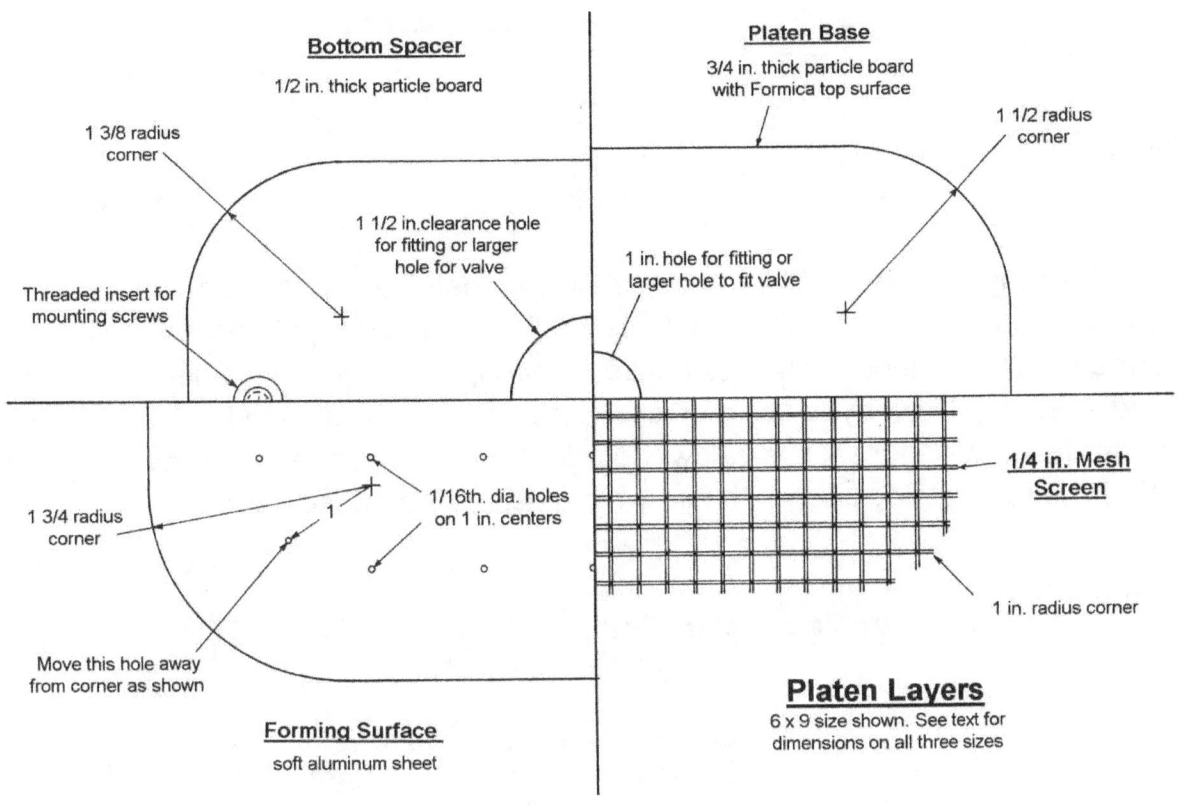

Bottom Spacer

1/2 in. thick particle board

Platen Base

3/4 in. thick particle board
with Formica top surface

1 3/8 radius
corner

1 1/2 in.clearance hole
for fitting or larger
hole for valve

1 in. hole for fitting or
larger hole to fit valve

1 1/2 radius
corner

Threaded insert for
mounting screws

1/4 in. Mesh
Screen

1 3/4 radius
corner

1/16th. dia. holes
on 1 in. centers

1 in. radius corner

Move this hole away
from corner as shown

Platen Layers

6 x 9 size shown. See text for
dimensions on all three sizes

Forming Surface

soft aluminum sheet

Selecting an Aluminum Sheet -

The aluminum top sheet needs to be soft enough that we can easily form the edges over with a hammer or mallet. The bottom sheet remains flat and the two will form an airtight sandwich.

You can start by looking for a suitable sheet anywhere cookware is sold. Look for a single layer cookie sheet without a rim if possible. These are usually thicker and soft enough to form easily. If done carefully, you can even use one with a non-stick coating but it's not necessary. Other aluminum baking pans or trays may work too. If you can't re-purpose some metal from a kitchen pan, you can buy new material from the source below. There are many types of aluminum alloys from soft to hard. Here is a list of common alloys if you decide to go shopping. The number in the center column is the Brinell hardness and lower means softer.

Aluminum Alloy	Brinell Hardness	Condition
6061-T0	30	Softest
3003-H14	40	—
5052-H32	60	—
6061-T6	95	Hardest

Ideally the aluminum thickness should be between 1/32 and 1/16th inch thick (1 to 1.5 mm) And it should be a soft alloy (30-40) that we can bend easily. The formed sheet will lay directly on the screen for support so it doesn't need to be very hard or strong. Don't try to use a steel sheet because it won't bend as easily. Based on this info, my suggestion is below. Note that if you search their part number, it will bring you to a catalog page with many other sizes if you need more.

McMaster Carr Company - Industrial Supplier
www.mcmaster.com

**Aluminum sheet, Alloy 3003-H14
.050 thickness, 12x24 in. Sheet
Part # 8973K477**

Forming the Aluminum Sheet

Bending the edges of the top sheet down, gives us a rounded area for the hot plastic to seal. Some people have trouble with this but its easy if you clamp it securely and go slow. Cut the aluminum sheet to the dimensions and corner radius shown on the chart. The aluminum sheet should be 1/4 in larger than the platen base all the way around. It's easier to lay out the perforated hole pattern on a one inch grid before forming the edges. Move the corner holes inward as shown to fit the curved corners and then lightly center punch the hole locations before drilling.

Use a scrap block of wood to clamp the aluminum sheet to the top side of the platen base. Center the sheet so it overhangs evenly on all sides. With the sheet clamped securely, use a smooth metal or plastic faced hammer and work your way around the platen, bending the aluminum edge down over the rounded wood corner. This looks harder than it is, just follow these tips.

Forming Tips:

-- A steel or hard plastic hammer works best, a rubber hammer is too soft. A steel hammer can leave marks so make sure the hammer face is smooth with rounded corners. If done correctly, there will be no marks on the aluminum sheet when you are done. Of course it will still work with marks, but do your best and maybe practice on some scrap pieces first or make the smaller platens for practice

– Take your time,.. You don't want to bend too much at once and stretch the plastic. Take many small taps close together just hard enough to make the metal yield. Go around many times and dwell in the corners where the metal must be shrunk to fit. The sides are just straight bends but material in the corners must actually shrink to make the bend.

-- Clamp the sheet securely so it doesn't slip. The scrap wood block on top should be 1/2 in. smaller all the way around than the wood base. The intent is to hold the sheet flat right up close to the bend area.

-- Use two or more C-clamps so you can move one as you progress around the platen and the other clamp will still hold alignment.

Drilling the Perforated Holes

If you haven't done it yet, the center of the sheet needs to have many small 1/16 in. holes drilled on the grid pattern you marked earlier. Lay the sheet on a flat surface and drill through from the back side with a sharp 1/16 drill bit. Use a high speed drill (1200 RPM or faster) and use a drop of oil for each hole to keep the drill from loading up. Aluminum drills very easily, but it will leave a nasty burr as it breaks through the top surface. Use a 1/4 in drill bit and spin it between your fingers to remove these burrs.

Glue it all together

Glue the fittings or check valve in place with adhesive and let them cure overnight. Clean the aluminum and Formica to remove all traces of oil so the adhesive sticks well. Use a solvent such as denatured alcohol, lacquer thinner, MEK or acetone. Don't use turpentine or mineral spirit's because these can leave a film.

Lay the formed aluminum top sheet upside down on a flat surface and place the screen inside it, making sure that it's centered. Apply a continuous 1/4 in dia.

bead of urethane or silicone adhesive just outside of the screen area. Make sure it's a uniform bead with no large bubbles or voids. Flip the wood base over and push it down into the formed top sheet. Be careful so the screen stays centered and the adhesive bead fit's around it. Push down and you should feel the glue spread out. Place a weight on top and leave it overnight to cure. The next day you can screw or glue the spacer board to the bottom side.

Attach the Platen to your Machine

The top of your cabinet will have a large clearance hole to access the vacuum fittings and smaller holes for the mounting screws. Lay the platen on top of the cabinet and swing the clamp frame over it. Center the platen inside the frame with shims if necessary. When you have the platen centered inside the clamp frame, drill the mounting holes up through the cabinet into the platen spacer. Be careful not to penetrate the sealed chamber between the aluminum sheets. Only two holes are needed per platen.

At this point you have two choices, if you don't anticipate changing platens often, then you can simply use wood screws to hold them in place. (drill pilot holes first) If you made several platens and want to be able to change them quickly or often then I suggest installing threaded inserts and threaded knobs. There are many types of inserts. so locate some first before drilling any holes. You can find these at many hardware stores. With the inserts installed you can then use 1/4-20 screws or a fancy knurled knob as shown in the drawing. Here are some examples.

Chapter 10 - Building the Oven

Sure, heating elements look simple and I already tried what you are thinking about right now. I cut apart toaster ovens, portable heaters, frying pans and griddles to find a good heating element. I tried all the easy stuff first, like heat lamps and heat guns, then I got catalogs of replacement oven elements and tried those. I came close with one from an electric bar-b-que, but it still cost more and didn't work as well as the custom heating element in one of my early machines. Call it a happy accident but it sent me on a mission to learn more.

It turns out that heat is a complicated thing and plastic prefers a certain kind of heat. All electric elements, Quartz tube, Ceramic, Cal-Rod, or bare coils have the same kind of nichrome resistance wire inside. Then I learned about heating element design. Slowly I came to grasp the reasons why all those small appliance elements didn't work as well. Without getting technical, heating element design is a deep topic and kind of fussy. It involves things like, convection, watt density and infra-red wavelength. All materials absorb infra-red heat differently. An element tuned for plastics will work better than one tuned for cooking food or heating a room. They all get hot but put out different infra-red wavelengths.

In addition to being tuned for plastics, the oven kit I sell is the cheapest way to go, and it's a proven design. You can spend a lot more money on other methods, but for the simple job of heating a plastic sheet, none of them will do any better than this. I did all this work to offer the cheapest most effective element to support my plans. Sadly, some people assume I'm keeping design details secret to force them to buy from me. The truth is that even slight material substitutions will affect the infra-red output. Selling kit's is the only way I can guarantee you got the best infra-red output.

After Covid and the supply shortages, materials for my kit's became hard to get. It's better now but I am motivated to find an alternative to my heating element kit as a second best option for builders in case the supply dries up again..

I will start this chapter showing how to build and install the heating element kit's I sell. A rational person will trust that this is the best and cheapest option as long as my kit's are available. **Chapter 11** will offer a second best option in case I ever have to stop producing my kit's or you just want to do it all yourself.

Assemble the Heating Element Kit

The fully assembled heating element from one of my kit's is shown below. The thin high temp insulation board is very special because it allows us to fasten the resistance coil directly to the board. Also the coiled element has been optimized for best IR output. Simple, quick build, durable and great performance. The element will become the floor of the oven box and you will need metal sides to contain the heat. We also need venting to provide strategic cooling around and through the oven.

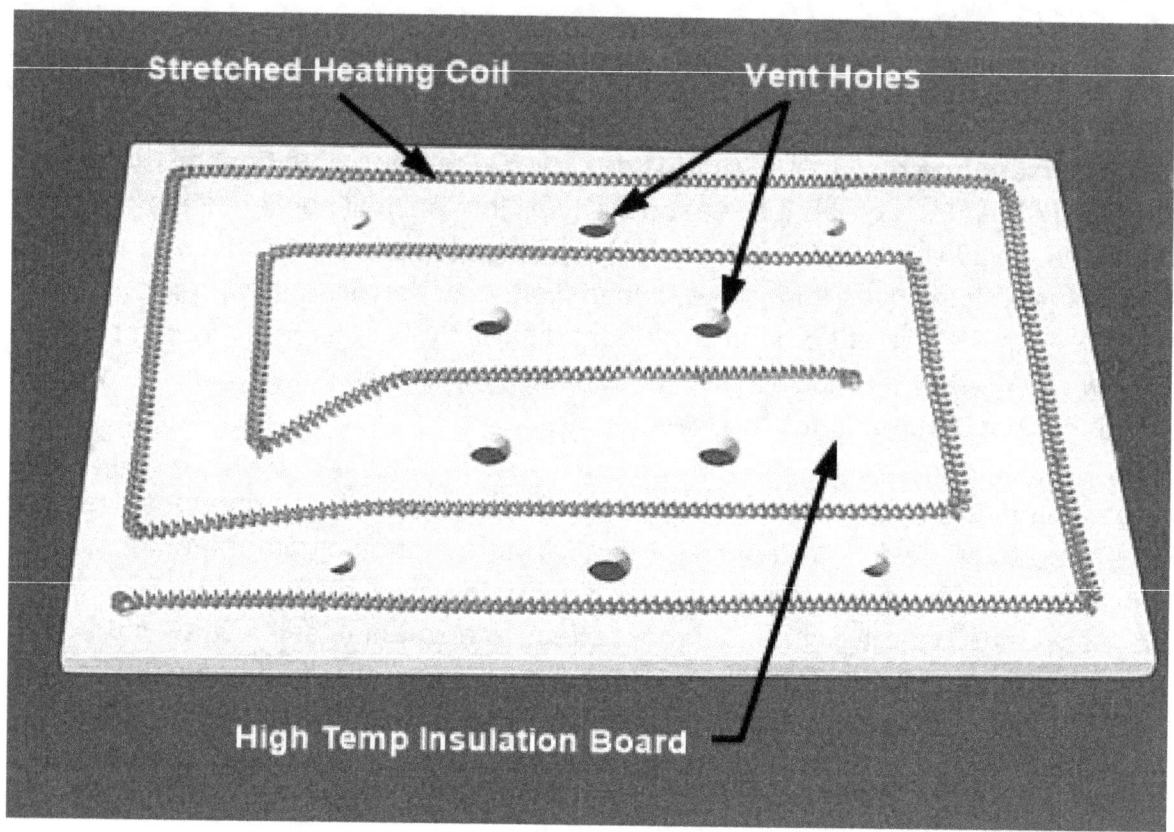

Stretched Heating Coil **Vent Holes**

High Temp Insulation Board

Kit Contents - The materials used for this kit are not commonly available. I have to buy large quantities and package them into kit's. Check www.Build-Stuff.com for availabilityand current pricing.

– You get one 12 x 18 inch sheet of high temp rated insulation board that can withstand direct mounting of the heating coil. This board is supplied a little bit undersized so it will drop into your oven box. I supply a drilling template for the holes.

– One length of pre-measured and coiled nichrome wire of the correct alloy, gauge and length, designed to operate on 115 volts. This element will provide infra-red radiation at the proper wavelength to match the absorption characteristics of the plastic. It will need to be stretched and fastened to the insulation board. Don't substitute any other type of wire.

– Stainless steel terminal screws, nuts and cotter pins to be used when fastening the coiled element to the insulation board. Do not substitute brass or steel fasteners. Brass corrodes and steel rusts

– Two high temperature connecting wires with special nickel plated ring terminals crimped to one end (6 inch and 12 inch). These wires are long enough to reach out of the hot oven area and connect with your wiring using the twist on wire nuts provided. Other types of wire and terminals will not withstand the heat, use only the items supplied in the kit.

Cautions

Do not operate this heating element in the shower... I was just checking to see if you were paying attention.

Wear safety glasses while reading this manual...... OK, now I'm sure.

This heating coil runs well over 1000 degrees F. and is both a shock and a burn hazard if used improperly. Do not use it without a safety screen in place to prevent any contact with the element. Always use a timer switch with a max time of 5 minutes.and never leave the machine unattended while being used.

Insulation Boards - The insulation boards that come with the kit's are a special high temp material that is not available in stores. Sometimes referred to as "millboard" it's an industrial material used in furnaces and other high temp uses. It simplifies the element because it can handle direct contact with a glowing red coil. It's also a low density material so it warms up fast and it's strong enough even when hot to avoid cracking or sagging. You can drill holes easily much like soft wood but avoid putting too much pressure on it because it can break. There may be printing on one side but either side can be used. You can make the inside of your metal oven box exactly 12 x 18 inches and the board is slightly undersized to drop in place.

Caution!

This insulation board has organic binders that will burn off during the first few uses giving off smoke and odor. Just provide good ventilation until this subsides. There will be some discoloration where the coil touches but it retains it's strength so don't worry about it.

When drilling holes, do it outdoors to avoid breathing the dust. Wear eye protection and wash your hands when done.

Do not get this material wet for any reason or it can swell up or sag and need replacement.

Drilling the Holes - Use the paper hole drilling template that came with the kit and center it over the insulation board. Poke through with a nail or something sharp to mark all of the hole locations. Use a 3/32 in. drill bit and drill the small holes all the way through. The two holes identified on the template as screw terminal holes should be drilled through with a 3/16 inch bit. The larger 3/4 in. and 1/2 inch dia. vent holes can be drilled with a hole saw or wood boring bit.

This material drills like soft wood or plaster. If you are using a wood boring bit for the vent holes, go slow and back it up with a scrap piece of wood so the drill doesn't take out a big chunk when it breaks through. You can also drill through half way, then flip and drill from the other side. Use a piece of sandpaper to knock off any ragged edges and loose bit's, then brush or blow off the dust outside and go wash your hands.

Stretching the heating coil - The coil supplied in the kit must be stretched before installation. Carefully form a loop on each end of the coil to fit around the stainless steel screws as shown. With a screw in each end you can stretch using a wood jig as described

Form the End Loops
Be careful not to nick or scratch the wire

 Good

 Bad

This loop is too long and will cut into itself when the screw is tightened

The center image shows a correctly formed loop. The last example to the right shows what you want to avoid.

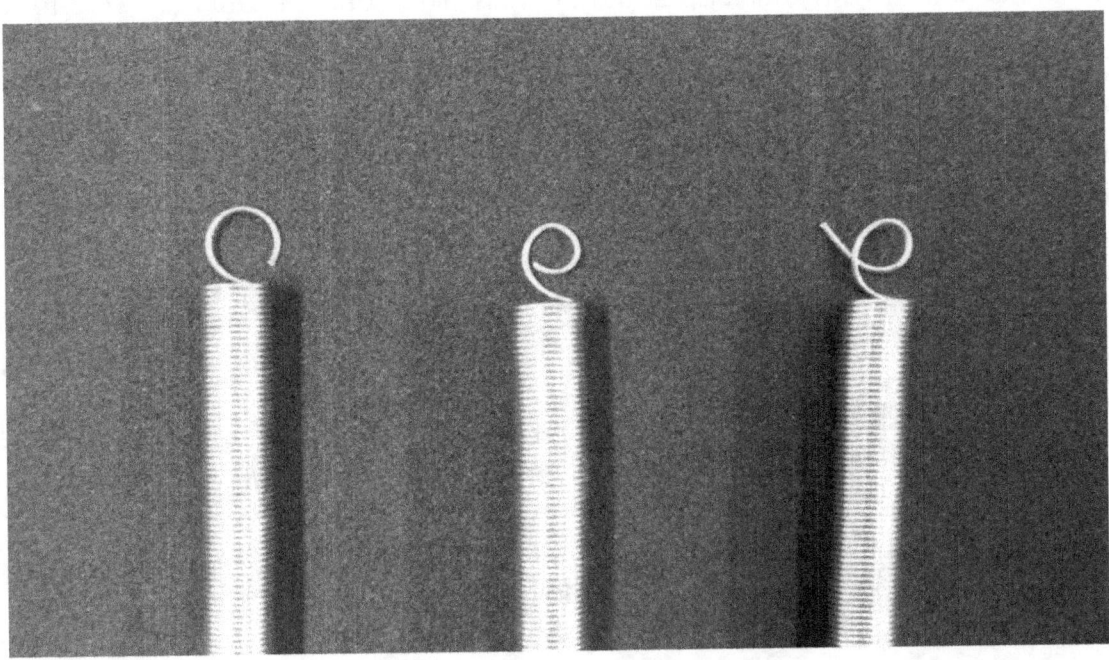

Be careful when you form the end loops or handle the wire, so you don't put any nicks in it. A nick causes a stress point and a hot spot that will lead to premature failure. Use only the supplied stainless steel fasteners for the terminals, never steel or brass.

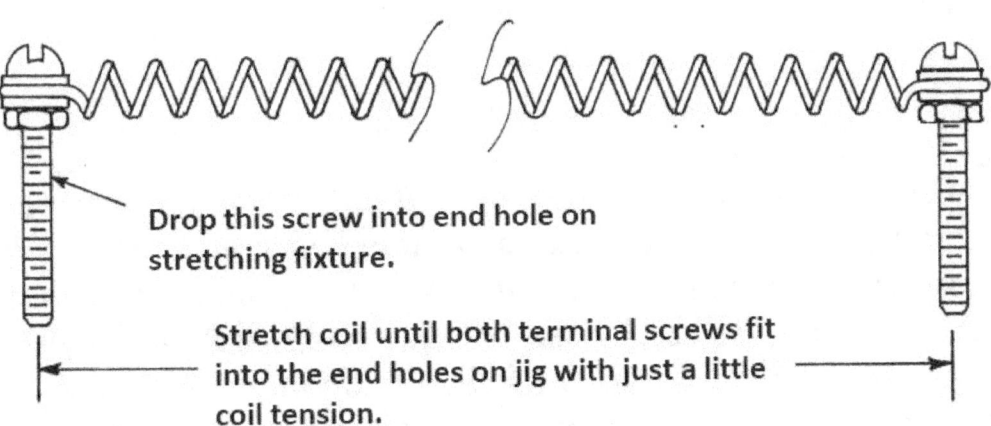

Drop this screw into end hole on stretching fixture.

Stretch coil until both terminal screws fit into the end holes on jig with just a little coil tension.

Stretching technique - The nichrome wire coils will need to be stretched, marked for corner locations, then fastened to the insulation board. You will need to make a stretching jig from a length of 1 x 2 lumber or any other scrap lumber you have that is at least 100 inches long. The idea is to start at one end and drill a 3/16th inch hole. Then measure and mark where the element will turn a corner as seen on the template and drive finishing nails so they stick up about an inch. Then drill the last hole 3/16 for the second terminal.

Start an inch from one end of the jig and drill a 3/16th. hole all the way through on the centerline. This will fit and hold one of the terminal screws. From there use a tape measure and make marks at the following dimensions in inches from the first hole.

16 inches - 26 - 42 - 50.5 - 64.5 - 70.5 - 82.5 - 87.5 - 90.25 - 98.25.

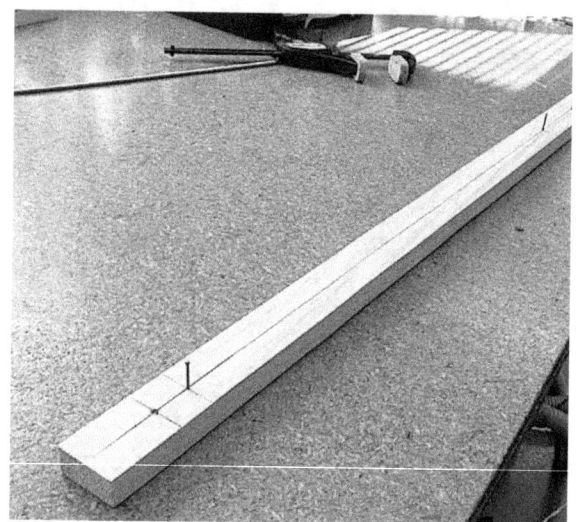

The last mark will get another 3/16th hole and everything in between will get a nail.

Clamp the stretching jig to a table so it won't move while pulling on the coil. Drop one terminal screw into an end hole and grasp the other screw then pull past the other end hole to allow for springback. It acts like a spring and will be too short when you release it. Keep pulling a little farther and check it each time until it's the correct length with light tension or no tension.

Tip:

The coil should stretch uniformly but only if nothing touches while stretching. Avoid the temptation to use your free hand to support or help the stretch. Also avoid having the coil drag against something while stretching. Let the stretch happen in free air as much as possible. Check the last six inches or so on each end. Sometimes the last few coils stretch more and you can see the difference in spacing. If this happens, do your best to massage it so the coil spacing is uniform. Over stretch as needed until the spring fits the jig with slight tension.

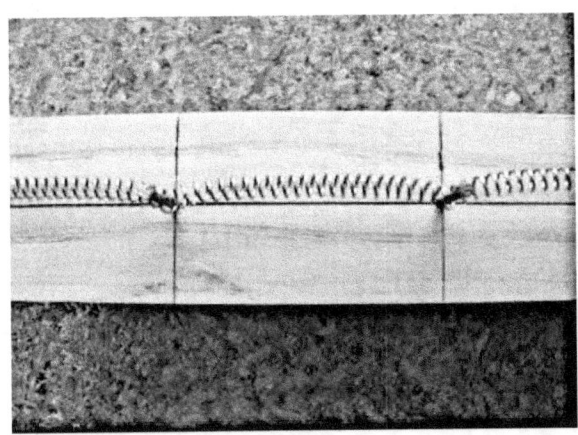
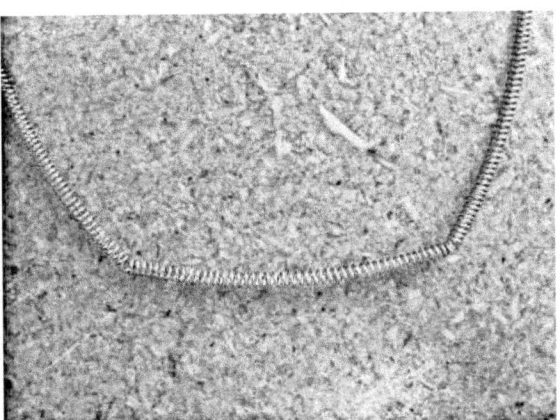

Now with the stretched coil still on the jig, use your fingers to gently press it against the nails as shown above. This just leaves a kink that you can see when the coil is removed. Use those kinks to help lay out the element when fastening to the board. Fasten the corners first then fasten wherever the coil crosses a hole.

Fasten the heating element to the insulation board with the supplied cotter pins as shown in the drawings. Start at one terminal and move from corner to corner. We will come back for the middle holes. It may help to spread the cotter pins open slightly and slip them over the wire and into each 3/32 hole you drilled earlier. Be careful to keep the coil straight and some tension between corners is OK so it's not loose enough to bulge between holes. When you reach the corners just bend the coil at right angles with your fingers.

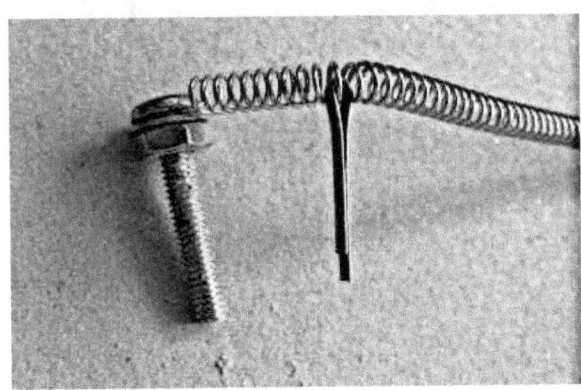
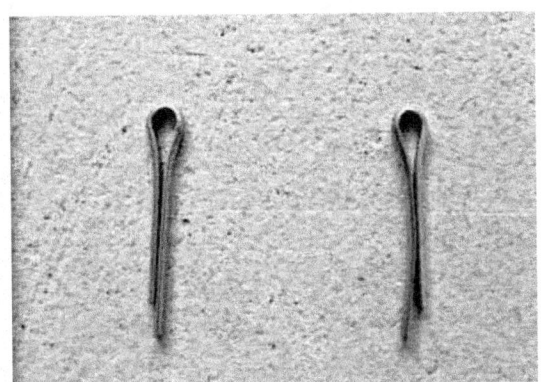

Tighten both terminal screws onto the wire loops securely and the threaded ends will pass through the insulation board and get double nutted on the bottom side to hold the ring terminals as shown.

Install Terminals
(shown full size)

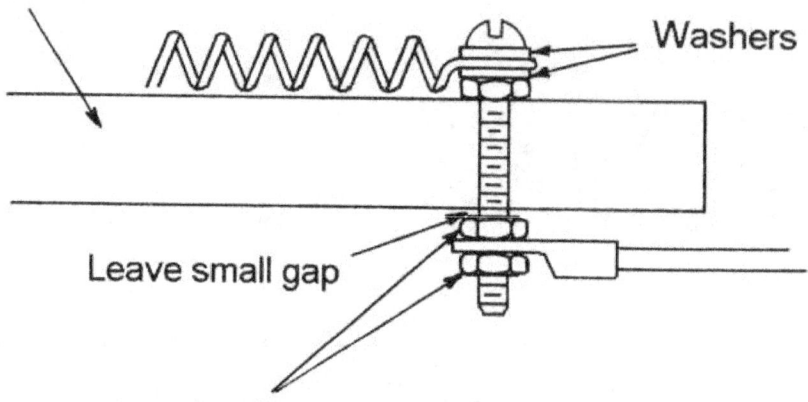

Insulation Board

Washers

Leave small gap

Use double nuts against wire terminal,
but don't clamp the insulation board

Caution

The screws should not clamp down on the fragile insulation board, the double nuts on the bottom should clamp the wire terminal tightly but still allow a loose fit through the board. The electrical connection is maintained but without stressing the soft board.

Your finished heating element should look like this and will become the floor of your oven. The round vent holes let cooler air in from underneath to replace the hot air exiting around the plastic above it. This effectively cools the center long enough for the corners to heat up and reach forming temperature.

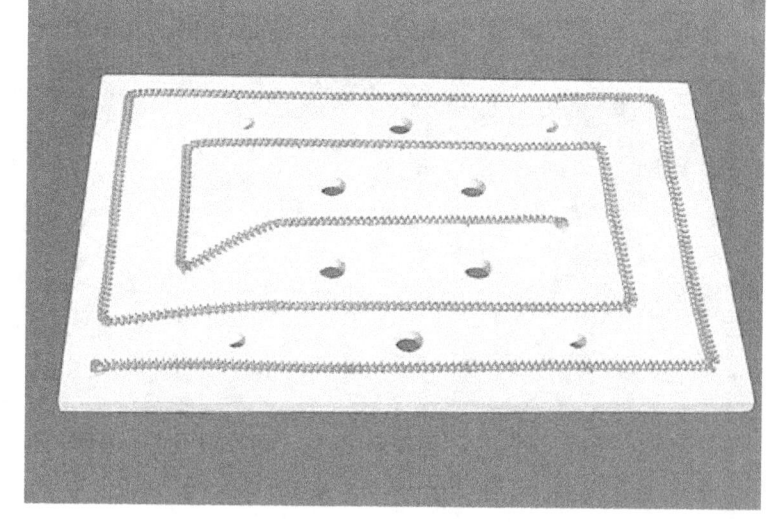

A sheet metal oven box is needed to reflect heat to the plastic and act as a barrier to keep the wood cabinet cool. We will use common galvanized steel sheet metal which is used for heating ducts in homes and is available from most large building centers. Any company that installs furnace ducts may be able to make these simple parts for you or possibly give you free scraps to work with. Any thin non rusting metal will work such as stainless, galvanized or plated steel. You can even cut up some cheap cookie sheets to use the metal. A last resort might be bare steel painted with a silver high temp paint to stop it from rusting. The thickness is not critical and lighter is better than heavier because it warms up faster.

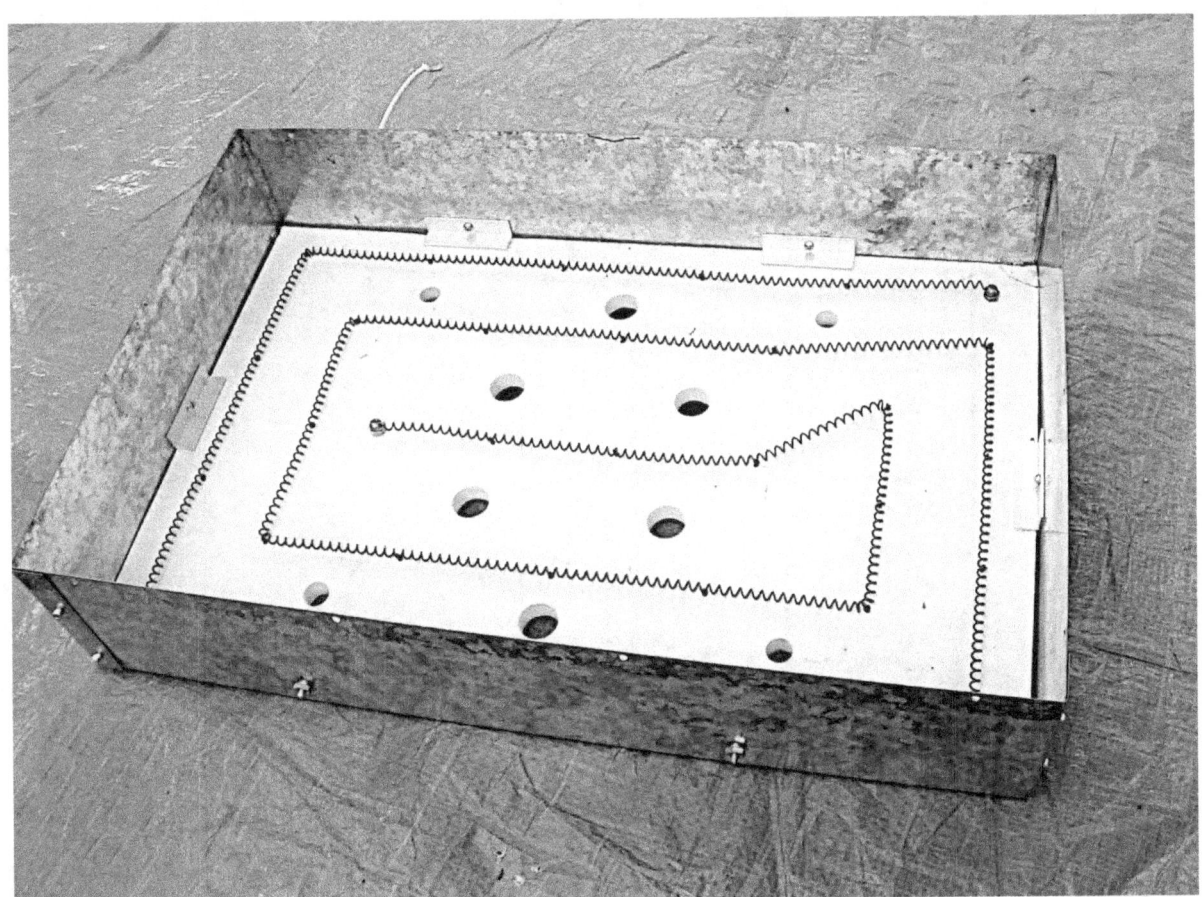

The heating element board will rest on the 3/8 flanges and be secured by six 2 inch long brackets made from pieces of aluminum angle as shown. These brackets should be attached to the sides of the oven box with small screws and nuts so they can be removed if needed. One of these screws can also be used to attach the ground wire described in a later chapter. Use lock washers under the nuts, if one of these screws came loose, it could touch the heating element and cause a short circuit.

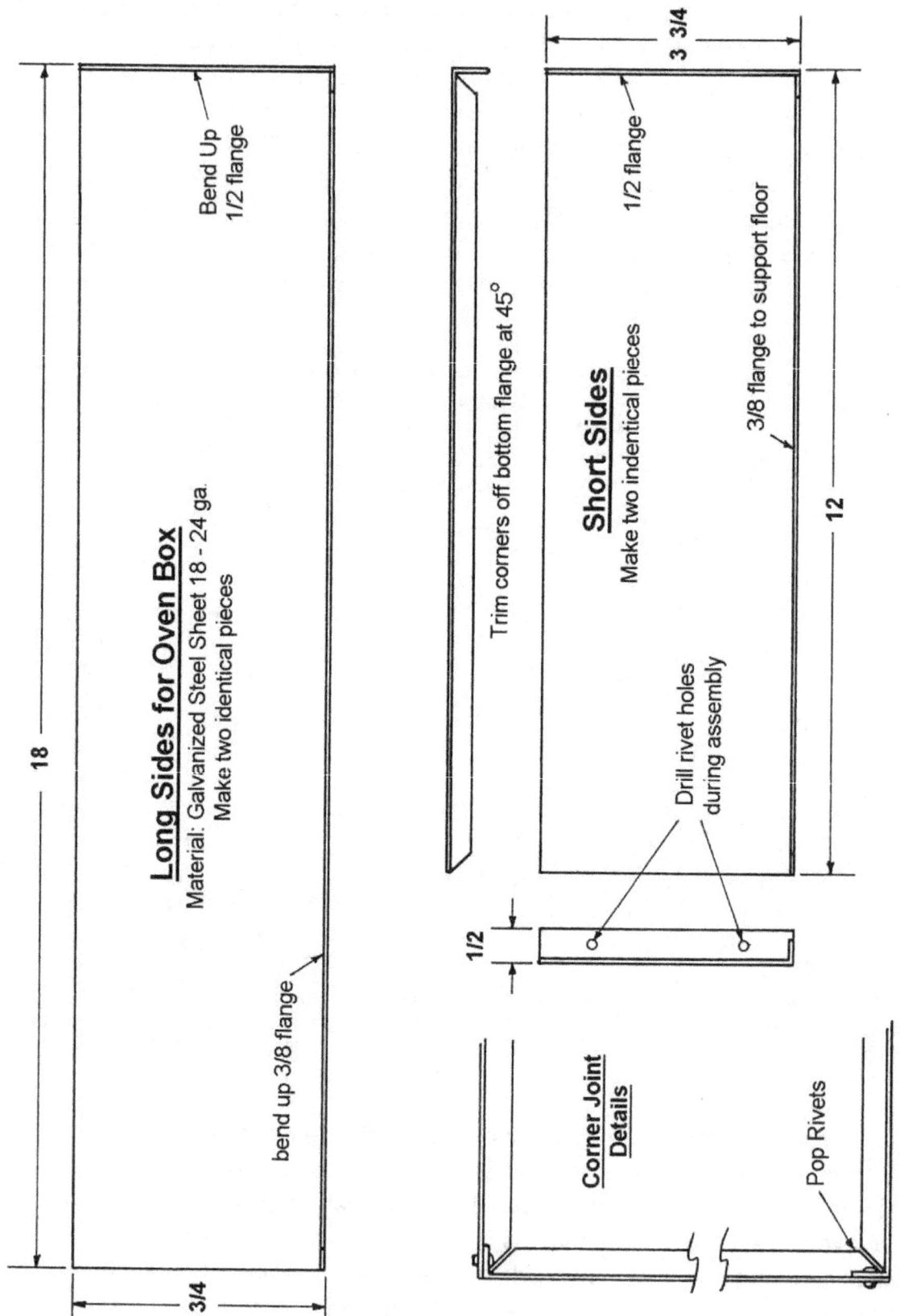

Long Sides for Oven Box

Material: Galvanized Steel Sheet 18 - 24 ga.
Make two identical pieces

Bend Up
1/2 flange

bend up 3/8 flange

18

3 3/4

Short Sides

Make two indentical pieces

1/2 flange

3/8 flange to support floor

3 3/4

12

Trim corners off bottom flange at 45°

1/2

Drill rivet holes
during assembly

Corner Joint
Details

Pop Rivets

The sheet metal oven box is formed from four pieces of metal and the corners create a slip joint. This will allow you to adjust the size of the box slightly. I suggest that you build the wood cabinet first and then fit the oven box into it. To form the flanges you can clamp the metal parts between two pieces of wood and use a hammer with many light blows. The bottom flange should be at least 3/8 in. wide to support the insulation board, but not longer than 1/2 in. or it may contact one of the cotter pins and cause a short. The sheet thickness is not critical, just use what you can find. The corners can be fastened together with ordinary pop rivets or machine screws and nuts.

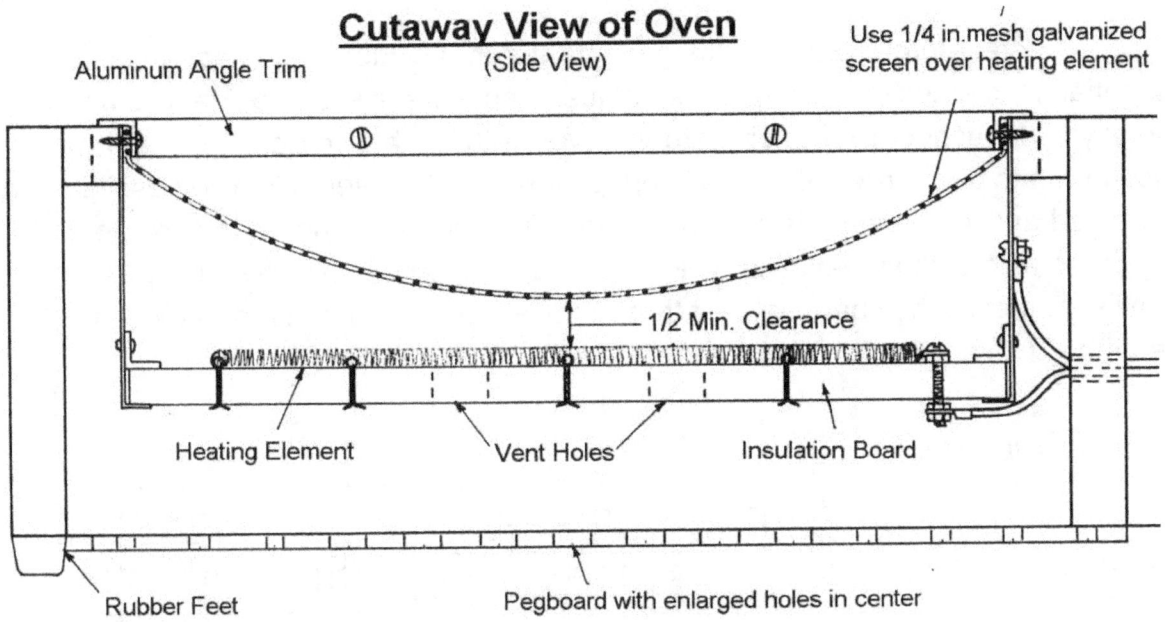

Cutaway View of Oven
(Side View)

Aluminum Angle Trim

Use 1/4 in.mesh galvanized screen over heating element

1/2 Min. Clearance

Heating Element   Vent Holes   Insulation Board

Rubber Feet

Pegboard with enlarged holes in center

Venting the oven is important for two different reasons. We need air venting around it to keep the wood cabinet cool. You can see that air can enter through the pegboard bottom and flow up and around the metal box. It then exits out the top vent slots in the cabinet.

The vent holes in the oven board itself are equally important. Hot air rises aggressively and will exit around the sheet causing a slight vacuum in the center. Cooler air needs to rush in somewhere to replace the exiting hot air. This can give you a cold spot around the edge wherever that happens. The vents allow it to pull cooler replacement air from under the element so it exits evenly all around the perimeter of the sheet with no backflow. The hole pattern and sizes are tailored so airflow will cool the center of the sheet and allow more time for the corners to catch up. This is a clever way to cool the hot spot in the center for more even heating.

Notice the wires inside the hot zone. There are three in total and two of them are connected to the element terminal screws and have a high temp insulation. You can only see one of those in the cutaway since its cut in half. There is also a ground wire screwed to the side of the oven box which grounds the box and screen to prevent shocks. The ground wire can be bare copper 14 gauge. The assembled oven board is secured with the six aluminum angle brackets.

Fitting the Safety Screen

The screen protects the heating element from accidental contact and is essential. Remember, the element is not just a hot wire, it's also a live wire and will shock you if touched. Do not operate the oven without this screen in place. It's not uncommon to drop one of the small spring clips into the oven. It's also possible to get distracted and let the soft plastic sag too far. The screen is curved to allow room for sag but to also catch the plastic. We will use a 1/4 inch mesh screen because the small spring clips may be able to fit through a larger ½ inch mesh and touch the elements. The screen is held in place under the aluminum angle trim.

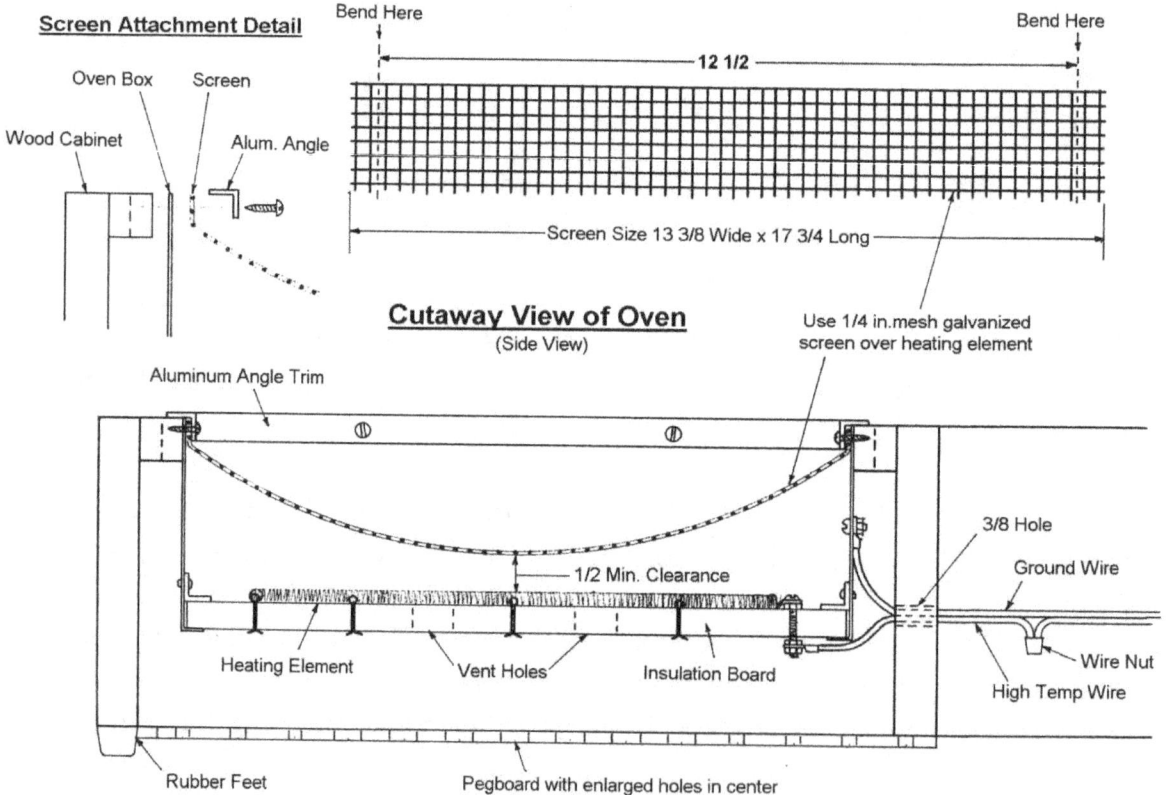

This type of screen is widely available in hardware and lumber stores. It is sometimes called "hardware cloth" and it cut's easily with sheet metal snips or a Dremel tool. I suggest that you cut a 2 inch wide test strip and experiment so you can verify the correct length. We want it to form a nice curve and have at least 1/2 inch clearance to the heating element as shown, but I would try for 3/4 to 1 inch. just to be safe. Bend up a flange so the screen fits neatly behind the perimeter aluminum trim piece.

Caution!

At least one wire of the mesh must go above the trim screws so the screen can never drop down onto the element, even if the screws loosen.

Add a Ground Wire - This is very important! In the unlikely event that the heating element wire should ever touch the oven box or screen, it would energize these parts and give you a shock if you should touch them. By simply adding a ground wire, a short circuit will now occur, causing the fuse or breaker to blow. The ground wire should attach anywhere on the sheet metal oven box. Ground wires are usually green colored or bare copper. You can use a crimp on terminal and attach it to one of the angle bracket screws or drill another hole and fasten it with a small machine screw and lock washer. Make sure this connection is reliable.

Final Inspection

Remember, the heating element is a "live" wire and so are the cotter pins underneath the oven. Check the bottom side and be sure that none of the bent over legs can possibly touch each other or anything else. Check the top side of the oven to see if the element is laying nice and straight and is not able to touch the aluminum brackets. Verify that there is at least 1/2 inch. clearance between the screen and the heating element even if you push down firmly. Never operate the oven without the pegboard cover over the bottom, and finally make sure the ground wire is installed.

Chapter 11 - Alternate heating elements

Why do we even need an alternate? The short answer is that I don't want to be the only source for optimized heating elements. I think I'm the best source but if my kit's ever become unavailable or you just prefer to DIY the whole project for the sake of learning, you need more choices. I don't even want to be a heating element manufacturer. For nearly 30 years I have been selling books, plans and machines and the heating element is a big part of what makes them work so well. There are many expensive and less effective ways to use re-purposed appliance or heater elements but none have worked as well as a simple tuned coil fastened to a high temp board. Both of the materials I use in my kits,.. the wire and board are specialized with enough nuances to make them hard to get and difficult to pair up correctly.

This chapter will offer help in finding substitutes that can work by selecting the right heating coils and suggesting ways to more safely use a common tile backer board material. It will even have about the same performance and cost as my kit's. The negatives are that the elements will take a little more time to build, will heat up a little slower and will likely have shorter life expectancy. The advantage is that if you can't get my kit's, these will work pretty well.

A match made in Heaven - or maybe Hell because they do operate at over 1000 degrees. The point is that the wire and board materials work together and affect each other to create an infra-red output. This is a deep topic that can fill a book but I'll try to keep it short. Infra-red radiation is like magic because it heats objects not air by vibrating their molecules causing internal heat generation. It works best when the output wavelength matches the absorption characteristics of plastic. In plain english it takes less energy and soaks in better if you get it right. Infra-red heat penetrates plastic much faster than heating by conduction or convection and uses less energy.

Here are some rules and reasons why it's hard to tune the infra-red output.

- The plastic being heated prefers IR wavelengths in a certain range.
- The coil wire (emitter) temperature determines the wavelength so we need to focus on the wire coil temperature,.. **not the air or plastic temp.**
- It's hard to measure a thin wire because if you touch it with a thermocouple it cools and a temp gun can't focus on a thin wire so it captures surrounding objects too. In other words we can't see or measure the IR wavelength easily.

- External factors such as room temp, drafts, and the thermal properties of the board can cool the coil so you can't use watts or amp draw as an indicator of emitter temp.
- You can't just calculate watts or resistance to get the right temp because everything around it changes the coils temperature..
- You also need to select element materials that can live happily well over 1000 degrees fahrenheit. That includes the coil, board, mounting hardware, terminal screws and hookup wire.
- The reason I sell kit's is so I can control all of these variables better to maintain the tuning.

Finding resistance wire

This is a special Nickel/Chromium alloy that is sold through industrial suppliers only. Everyone just wants me to tell them the resistance or watts needed so they can use some old MIG welding wire, piano wire or guitar strings. It scares me what people have tried. If you see anything else but NiChrome wire used on an internet forum,... don't try it.!!! I won't get into all the resistance alloys available or the science of designing an element. I'll just give you some basic specifications and one source to get coiled resistance wire. I buy bulk wire and coil on my own machinery or buy coiled wire whenever I can. You can buy pre-coiled wire from the source I provide below but it's not pre-measured. You will have to carefully measure it's resistance with a meter and cut the correct length.

I can also sell you pre-measured coil kit's with stainless steel hardware a lot cheaper so call for availability. (248) 391-2974 www.build-stuff.com

Otherwise you can get it in 10 ft lenghts from.......

MOR Electric Heating Co.
Part # COILHD10004
www.infraredheaters.com
(800) 442-2581

The coil listed above is much too long so you will also have to measure the resistance as accurately as you can and cut the proper length. Because it's a tightly wound coil you will have to stretch it a little while measuring to keep the coils from touching each other. You can do this any way you want, but I use a simple jig with two dull knife blades fixed about 18 inches apart and connected to an ohm meter. Use trial and error until you find the length needed to measure exactly 8 ohms. Follow the instructions in Chapter 10 to stretch the coil, form the end loops and install it.

Resistance Coil Specifications.

- "Nichrome 60" resistance wire alloy
- 17 gauge wire, close coiled around a ¼ inch diameter arbor.
- Measure and cut to 8 ohms total resistance when slightly stretched to keep coils from touching. Then stretch to final length and install
- This will draw around 13.5 amps depending on actual voltage and a little over 1600 watts. OK to use on a 120 volt, 15 amp residential circuit.

This element will be optimized for use on a low density insulation board ½ inch thick or less with an area of 216 square inches or 12 x 18 inches. If the insulation board is too thick or dense it can draw too much heat away and cause the wavelength to be off. It is designed for 120 volts and is meant to be the largest "comfortable" load able to run on a standard residential 15 amp outlet with no other large loads. Don't try to spread it out over a larger or smaller area. .

Warnings!

Never run this element on 220 volts! Be aware that any other wire type, gauge or alloy will change your tuning and can become dangerous. Don't substitute any wire that doesn't match these specs, and don't change the total element area, or venting, or try to add insulation. All those things change the output. Only run these elements on a 120 volt 15 amps (or larger) circuit that has no other large loads on it. Don't use it in the rain,.. Don't sit on it,.. or touch it with your tongue and most importantly......

NEVER leave a hot element unattended!

Hardware required

You will need stainless steel hardware to mount and connect this element. Brass or plated steel will oxidize at high temps and bare steel will rust. These items are all included in our kit's. If you decide to source it all yourself, read on for the info you need. You might first try your local hardware store for the nuts and screws because the same items are usually only available in full box quantities online. The coiled nichrome is the same way. You can order it online but you will pay more and get much more length than you need.

These items are available from McMaster Carr Co. at www.mcmaster.com

Terminal screws - You can use regular round head screws but these are called "binding head" screws and have bigger heads meant for electrical terminals. Part # 91793A201, 8-32 thread stainless steel 1 ¼ inch long, 2 required.

Nuts - Stainless steel 8-32 thread, Part # 91841A001, Qty. 30 (use for spacers between boards too)

Washers - Stainless #8 washers Part # 92141A009, Qty. 4

Cotter pins - Stainless 1/16" x 1" Long Part # 98401A417 Qty 35

Wire Terminals - Non Insulated High-Temperature Ring Terminals, 900°F Max Temperature, for 12-10 Gauge and # 10 Screw Part # 69405K57 Qty. 2

High temp wire - 12 Gauge 250C (482F) Silicone Rubber Wire. # WIREHT10008 Sold By The Foot from https://morelectricheating.com Qty. 5 ft. This wire supplies power to the heating coil and can handle the high temps directly under the oven.

Alternate Insulation Boards

     If you choose not to use my element kits, you can try to find the same industrial high temp boards I use from the sources below. The final option might be to use a common tile backer board. First let's understand what the board needs to do for us. This is the foundation of our oven and is the hardest item to find. I have tried many materials over the years but the one I use in my kit's is often referred to as "Millboard". It will handle temps well over 1200 degrees. It also has low mass and density to warm up fast, because light and thin is best. We will use a ¼ inch thick board. Again, it's an industrial material not usually found in small quantities. Only a few sources exist and they don't sell small quantities directly. You need to find a distributor and buy full sheets which often must be crated and shipped by truck unless you find someone who will cut them.

     McMaster.com is one distributor for the same Millboard I provide in my kit's but again they are larger sheets and will likely cost around $120 just for the wood crate and shipping charges by truck. If you live near one of their warehouses this may be an option but if you have to ship it won't make sense.

     www.mcmaster.com/9362K13/ Rigid Millboard Insulation Sheet for Furnaces, Ultra-Thin, 1380°F Maximum Temperature, 1/4" Thick, 39" Wide, 39" Long. Prt # 9362K13

     There are other types such as ceramic fiber and calcium silicate that are a little more common and may be found in smaller sheets for replacements in wood burning stoves and fireplaces.Look for a density between 24-50lbs. These are a little more like plaster and can chip or break, ½ inch is the thinnest you will find so you will need longer cotter pins and terminal screws. Below is one such source, Google for more.

     www.simondstore.com Ceramic fiber board, 2300F size .5-inch x 12 x 24-inch

Tile backer cement boards

     So far all the materials mentioned above were rated for temps above 1200 degrees.There is another very common and inexpensive material that looks similar to millboard but is not temperature rated. It's sold in lumber stores for use as a tile backer board and is made from cement, sand and other fillers. It is fireproof, but more prone to degradation and cracking. One brand of tile backer board is often used by hobbyists and is popular on internet forums for use with vacuum forming ovens.

James Hardie Co. "HardiBacker" tile backer board. It comes in ¼ inch thick and is readily available in 3x5 ft. sheets. They also make similar materials for use as siding, underlayment and trim. They may all be the same composition but I don't know for sure so stick with the HardiBacker tile board. I would avoid all other variants and brands for reasons listed later.

Regular ceramic floor tiles are similar in the sense that they also won't burn but are even more likely to crack than tile backer. Dense ceramic also draws away more heat and affects IR tuning more. Large floor tiles can be made to work and are certainly cheap but overall I think they are a bad choice for performance and safety. I won't address them any further so experiment at your own risk.

The problem is that many people use HardiBacker and post about it but the reality is that it has a solid history of failure. People who post on forums rarely ever come back and edit their posts when it fails so this is not as well known. I have been following its usage for years but kept my distance out of concern for safety and have not mentioned it in my plans until now. I still don't endorse it and my feeling is that we should use the right materials whenever we can. After we had supply shortages I had to revisit this material as a possible or at least short term substitute. But, only if I was able to come up with a better method to use it that greatly reduces the tendency to crack and greatly minimizes the risk if it does. I will share my ideas on that.

Warning and Disclaimer!

HardiBacker board can function in an oven but its weaknesses should be addressed as described and you should expect less life from it. I don't recommend it and am only trying to suggest safer ways to use it for those willing to accept some risk. HardiBacker is popular and I can't prevent its use. I am only sharing ways to make it safer,.. You must decide if it's safe enough for you!

Here's why it fails - It's a harder and denser material than Millboard which makes it more brittle. The uneven and intense heating and cooling can often result in cracks. Its composition also doesn't seem closely controlled due to recycled content and it seems to vary from batch to batch. Usually a crack is harmless but if it's in the wrong place or very large it may collapse and cause an electrical short.

Here's how we reduce the risks - By cutting it into smaller segments and using two layers separated by flexible spacers we can let it expand and contract. Kind of like the control joints in sidewalks and driveways. The top layer in contact with the heating coils is now in six segments while the bottom sheet becomes shielded from much of the heat so it can remain in one piece. In addition there is an air space between layers with flexible spacers and vent holes to allow air cooling between layers. If the top segments crack, they are still held in place by the bottom layer and multiple fasteners.

What about other brands of tile backer boards? - I was going to make this a thorough review so I gathered samples and made up test elements then let them bake for a whole day. Right away I could rule out most of them. There are cement, gypsum and foam based tile backers. Foam will burn so avoid those. Gypsum won't burn and it's easier to cut than cement but it is weakened too much by the heat so avoid those too.

Cement based boards are the only ones you should consider but they come in two flavors. There are rough textured boards with fiberglass mesh skins and rounded edges. These are too crumbly and brittle. The other is a mix of portland cement with sand and mystery fibers,.. Possibly fiberglass? For the purpose of these plans I'm going to limit discussion to only one brand and product that is mentioned the most on forums. **"James Hardie" brand "HardiBacker"** It's identified by the embossed name and grid lines on one side. The back side is smooth but with a whiter color and more mottled appearance.

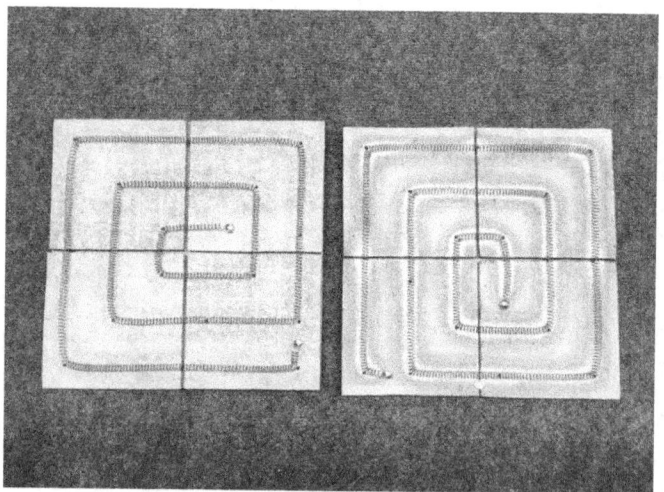

These two test elements were left running for a long time to check for cracking with the two layer construction. The one on the right is a Gypsum based tile backer and it was scorched and weakened enough to fail this test.

The left sample was HardiBacker and its appearance and strength were unchanged. I mounted the coil on the back (smooth) side of the board.

This torture test ruled out the Durock brand gypsum based boards and also confirmed that the two layer segmented construction does resist expansion cracks well.

Cutting and drilling HardiBacker board is well documented on the web so I'll be brief here. The manufacturer says you should score and snap it to avoid making dangerous silica dust. Installers also use carbide tipped wood blades or diamond grit blades to saw it. You can drill the holes with carbide tipped masonry or tile bits. Visit the tile section of any store to find many round and straight saw blades as well as drill bits for tile work. Let's get started.

These boards will drop into the sheet metal oven box described in Chapter 10 which has an inside dimension of 12 x 18 inches. The board therefore, must be made 12 x 18 inches or slightly smaller so it fits inside. I have the tools so I cut mine with a 7 inch diamond grit tile blade on a table saw. You can also find a 4 inch blade to fit an angle grinder and follow a cut line manually. A diamond grit straight blade in a jig saw will also work great. To avoid dust you can score and snap the board but do a test cut first to see how it goes. It's hard to be accurate with this method. You can buy a carbide scoring tool just for this. A normal razor knife will work well too but is tough on blades.

I think a diamond grit blade in a jigsaw is a good compromise of accuracy vs cost. Beware that silica dust is a very bad thing to breathe so take all precautions and wear safety glasses.

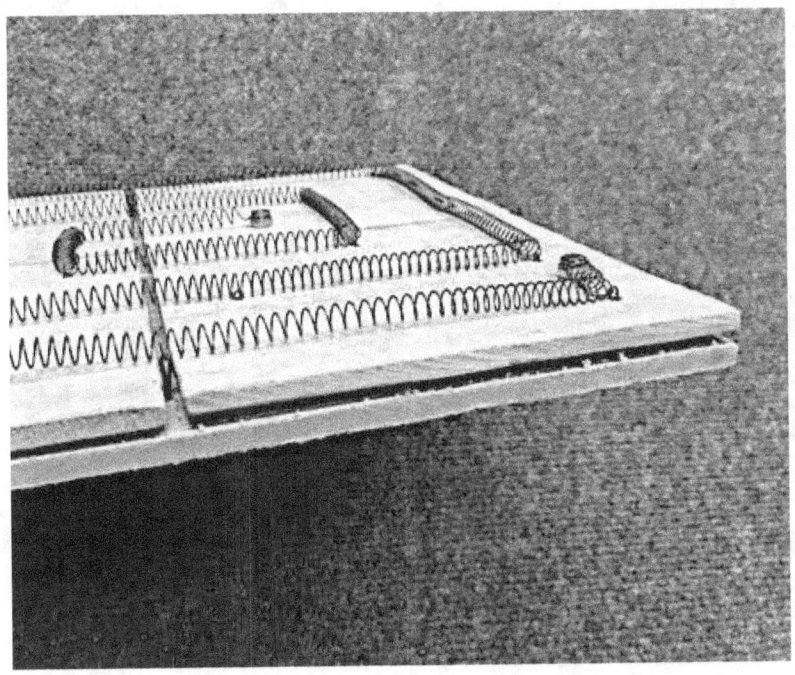

The two layer construction can be seen on this test element. The top surface that touches the coil is made from smaller segments with gaps for expansion. The bottom layer is one piece and is shielded from the heat but it also has an air gap. All the pieces are held together with a flexible adhesive and spacers as well as the terminal screws and cotter pins holding it all together. It will be much harder to crack, and if it does, the pieces are held in safely together.

Tip:

The lifespan of HardiBacker in this application is unknown at this point so check periodically for cracks or degradation from heat. Check the bottom side as well since it's hidden from view. If any cracked parts are free to move excessively or fall out, you will have to replace the boards. The worst case condition for cracks is to start the oven up in a cold environment which could cause thermal shock. These machines need to be used in still air in a warm room for best results anyway. For a longer life try not to start up the element in a very cold place.

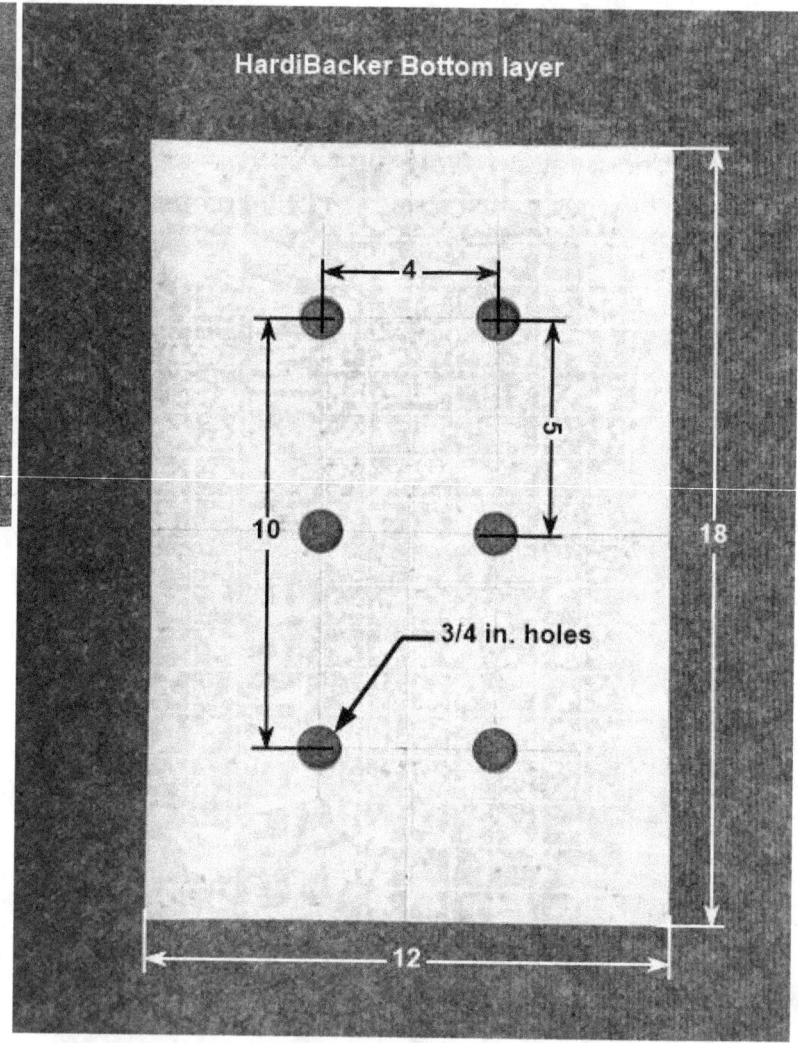

Cut the bottom board to 12 x 18 in. so it fits into the sheet metal oven box. The six smaller squares should be a little undersized. 5 15/16th squares are good to allow for a small expansion gap between them. Before we glue the boards together with spacers, we need to drill some ¾ in. vent holes through the bottom board in the locations shown. This serves two purposes, It lets air flow between the layers and up through the top and cools the center of the plastic slightly to give the cooler corners more time to heat.

The HardiBacker material is cement so it's hard to drill with woodworking bits or holesaws. Later you will need two small carbide tipped drill bits to drill the coil mounting holes. You can use the smallest one now to drill closely spaced holes inside a 3/4 inch circle. Knock out the center and clean up with a half round file or a grinding stone on a Dremel. These vent holes through the bottom board will be hidden by the top layer so it doesn't matter if they look a little rough. They do need to be located where shown to miss all the other holes later.

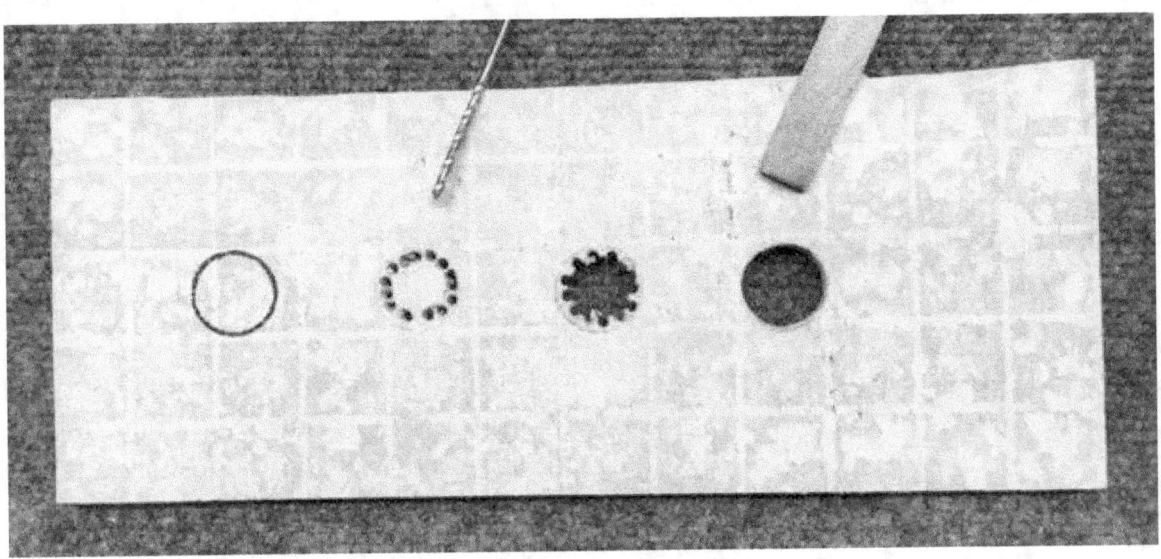

The photo above shows how to make big holes with a little drill bit. Once you get the vent holes done, let's assemble the layers.

Why two layers? - The top layer will be smaller squares with expansion gaps between them to reduce or eliminate cracking.It also shields the bottom layer so it can survive as one piece without cracking. With all the parts and nearly 40 holes, it takes some planning to make sure everything misses everything else. The two layers are held

together with an air gap between them. Each top square will have four spots of RTV silicone rubber with a small machine screw nut embedded inside to act as a spacer. The terminal screws and all the cotter pins holding the coil will also hold the layers together.
Silicone rubber is a high temp adhesive/sealant but there are many varieties. This one is sold for automotive applications and has an extra high temperature resistance. It can be found at many hardware and auto parts stores.
Permatex.com Ultra Red RTV Silicone Gasket Maker
Part# 81630 3 ounce tube.

Assemble the two layers together

Mark the four corner locations where we will use adhesive spots to hold the layers together. We can use some small 8-32 machine screw nuts for spacers to keep the layers apart. They will need to be placed accurately to avoid all the other fasteners we will use later

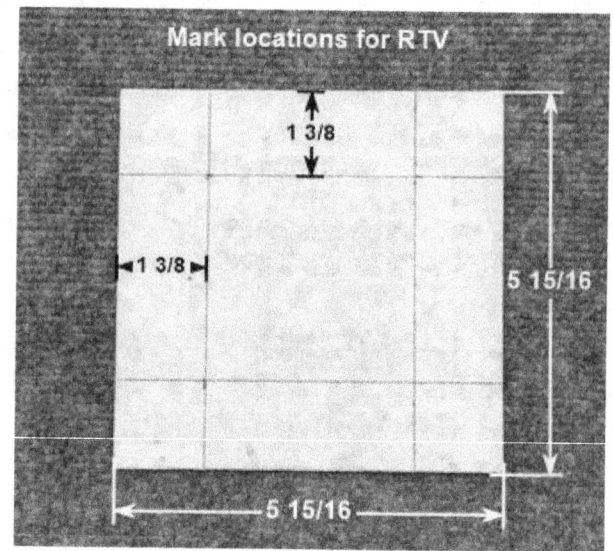

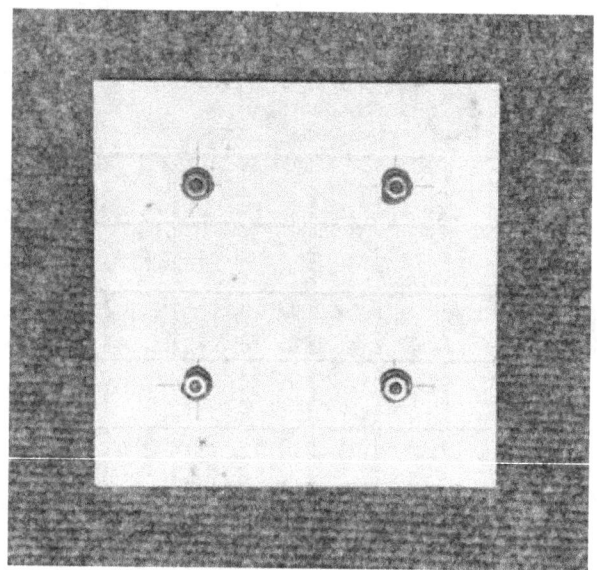

The first step ist to squeeze a spot of high temp RTV silicone at each corner location you marked and carefully push a small nut into each spot. Be careful on the next steps so they don't get moved out of place.

Place a larger drop of Silicone over each nut. The resulting silicone spot should be a minimum of around a half inch diameter and half inch high. There is no harm if you use too much because it will just squeeze down until the small nuts act as spacers between the boards.

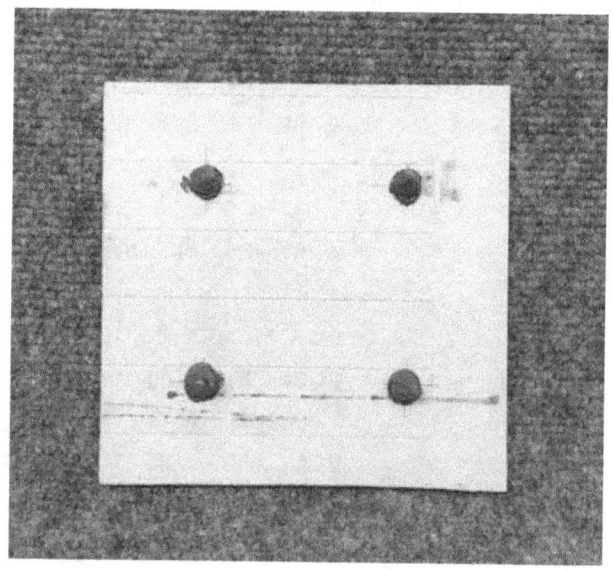

Carefully place each square onto the bottom board with outside edges flush. Try not to slide them around after placement. We want the small nuts to remain in their marked locations. Press down until you feel the silicone squeeze out the nuts control the spacing. Adjust as needed so the expansion gaps are uniform.

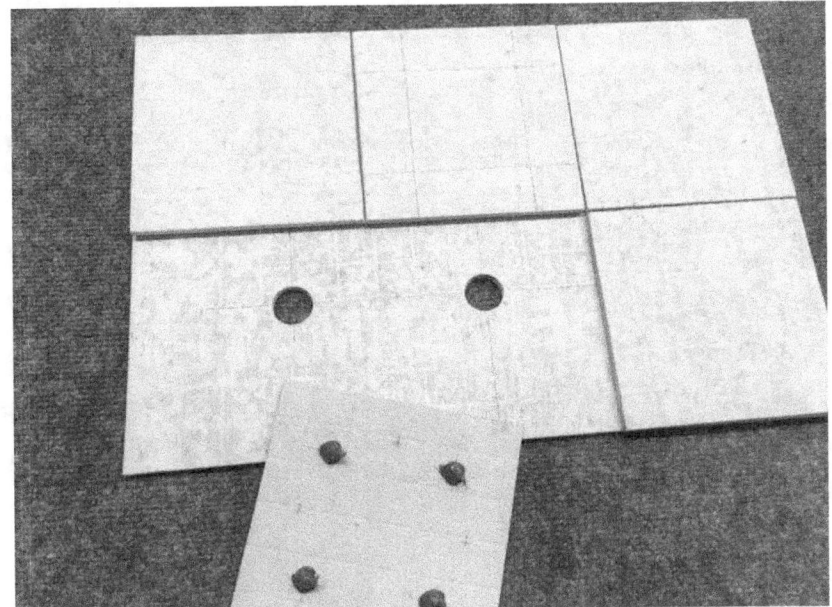

so

<u>Mounting the heating coil</u>

We will use the same heating coil and stretching jig we used for the kit elements Refer to chapter 10 for those details. If you purchased the coil from me it will be pre measured, otherwise you need to measure it and cut a length that measures exactly 8 ohms. Prepare the end loops, stretch on the jig and mark the corner bends. The kits came with a drilling template but this two layer HardiBacker board will need a similar but different hole pattern as shown below.

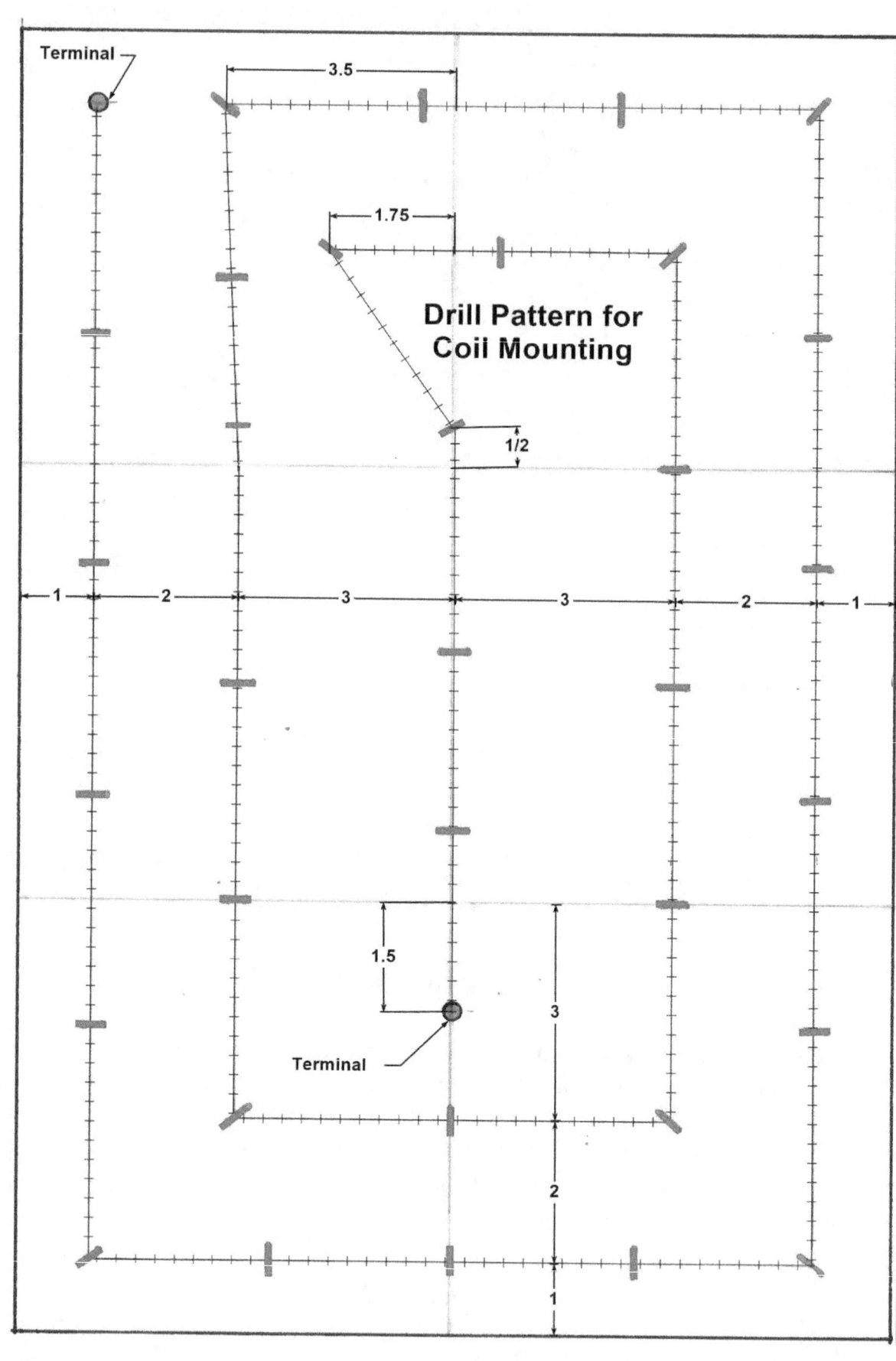

Drill Pattern for Coil Mounting

Drill Hardi board and mount coil

The layout drawing shows the pattern for mounting the coil. The easiest method is to draw a rectangle one inch inside the board's edges, then draw another rectangle two more inches inside the first one. That pretty much marks the coils path as it spirals in from the top left corner and terminates on the centerline. Drill two 3/16th holes with a carbide tipped masonry or tile drill at the two end points marked "Terminal" The red marks along the path represent cotter pins used to hold the coils to the boards. Drill ⅛ inch holes at each corner from terminal to terminal and then wherever the red marks are in between. You should end up with a 3/16th hole at each end and thirty three ⅛ in. holes along the coils path. Drill with light pressure especially when breaking through and blow out all the dust.

Caution!

The first time you use the oven it may smoke heavily as the organic binders burn off. This is harmless and will not continue beyond the first few uses, Just provide ventilation until it stops

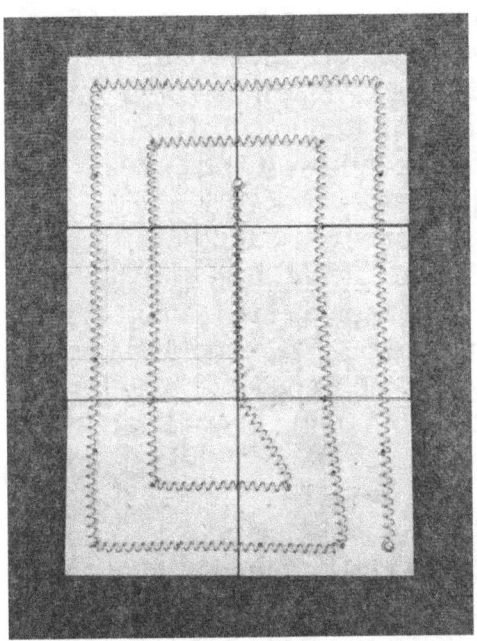

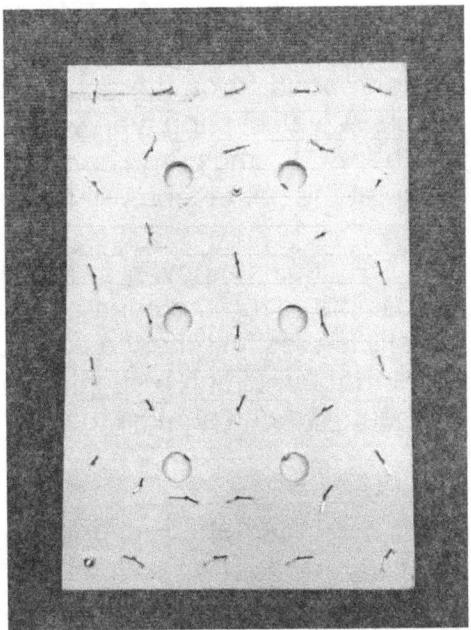

Top view on left
Shows element ready to install into oven box. You can see the expansion gaps between tiles.

Bottom on right
You can see the vent holes and bent cotter pins. Trim any cotter pin ends if they come too close to the metal oven box.

Chapter 12 - Electrical Wiring

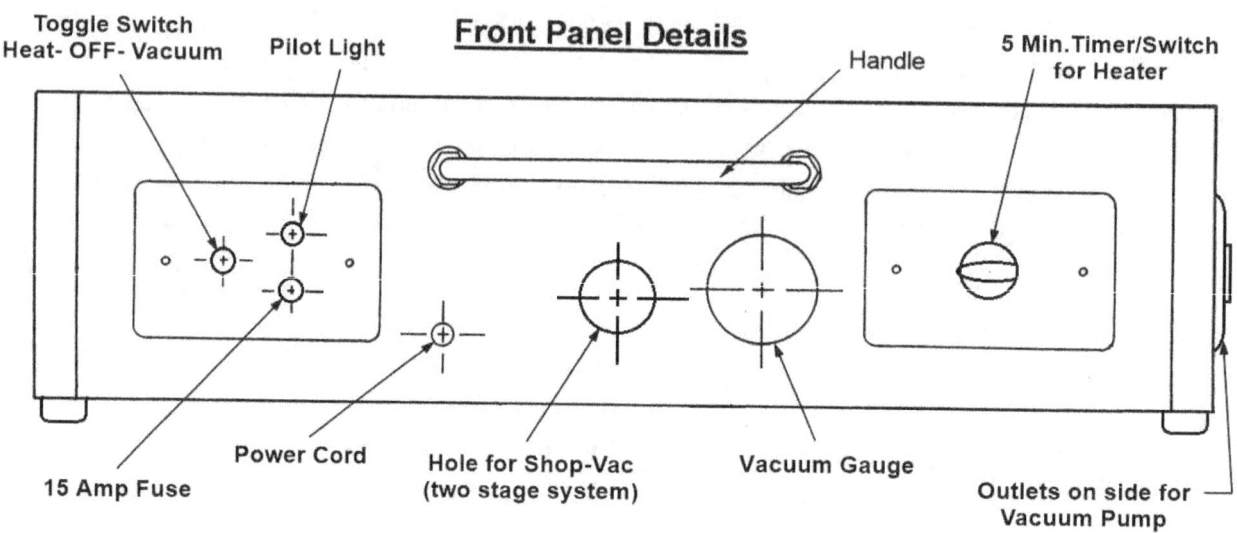

Front Panel Details

Toggle Switch
Heat- OFF- Vacuum

Pilot Light

Handle

5 Min.Timer/Switch
for Heater

Power Cord

Hole for Shop-Vac
(two stage system)

Vacuum Gauge

15 Amp Fuse

Outlets on side for
Vacuum Pump

Safety First... We want to manage our two big loads (**heat** and **vacuum**) so it will work on a standard residential house circuit. Since we are limited to a total of 15 amps we have to put some safeguards in place. The oven itself draws around 13.5 amps depending on the actual voltage. This is the largest oven we can use on a house circuit. A typical shop type vacuum cleaner or an electric vacuum pump will draw another 8-10 amps. You can see that any attempt to run both the oven and vacuum pump at the same time will exceed 15 amps and blow the fuse.

By using a power switch that has three positions, **Heat-Off-Vacuum**, It becomes impossible to run the oven and pumps at the same time. That solves the load problem but still leaves the chance you could walk away and leave the oven running. We can solve that by adding a simple 5 minute timer switch between the panel switch and element. We can set this to control the heat cycle and "time out" if the oven is left on too long.

It's important to understand that this machine will need almost all of a 15 amp. circuit by itself. This means that no other significant loads should be plugged into the same circuit. I suggest doing a little detective work to verify this. Use a lamp to check the outlets in the area and have someone watch it while you turn off the circuit breakers in your main panel. Unplug or turn off any other large loads on that circuit.

Lets start by mounting all the electrical components in the wood cabinet so we can connect them with wire.

Electrical Components

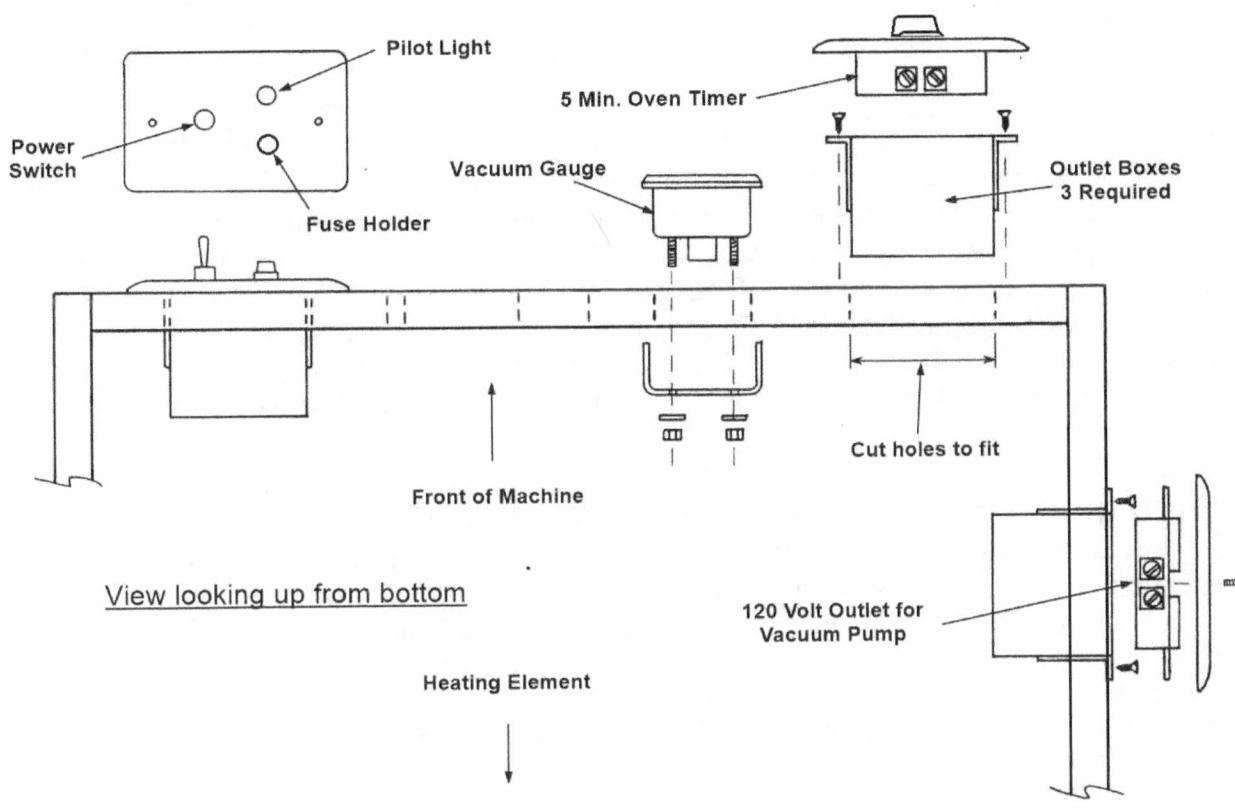

Pilot Light

5 Min. Oven Timer

Power Switch

Vacuum Gauge

Fuse Holder

Outlet Boxes 3 Required

Cut holes to fit

Front of Machine

View looking up from bottom

120 Volt Outlet for Vacuum Pump

Heating Element

Wiring - Use only 14 gauge or heavier wire for this machine. The pilot light is the only item that can use lighter wires (22 gauge min.) because it only draws a small amount of current. The heating elements will need high temp wire inside the hot zone which Is included in my kits and specified in the previous chapter. That high temp wire can join to THHN wire once outside the oven area THHN is standard residential house wire available at any hardware store. Typically we use black for hot, white for neutral and green for ground.

Safety Timer - This adjustable timer will automatically shut off your oven in 1 to 5 minutes. Just twist the knob to re-start it. This is protection against getting distracted and leaving the machine run while unattended. **https://www.mcmaster.com/7014K6/**

Power Cord - This must be at least a 14 gauge three conductor cord for use with three prong grounded outlets. 6 to 8 feet should be plenty of length.
McMaster Carr part # 70355K112

Power Switch - This must be an On-Off-On, double throw switch. It can have one pole (three terminals) but it's easier to find switches with two poles or six terminals and just use three of them. You know you have the right switch if it has three distinct positions with the center being off. Don't confuse this with identical looking switches that have only two positions This switch must be rated at 15 amps or more, this makes it a little difficult to find. Most hardware stores or lighting supply stores will have one, but double check to be sure it's at least 15 amp. rated. **McMaster Carr part # 7343K731**

Pilot Light - This is optional but a little hard to find. It's nice to let you know when the element is energized. It should be a bright, preferably red light. You will notice that any product with a heating element always has a pilot light to remind you it's on. This light should have an LED or neon bulb because they have an almost unlimited life span. An incandescent bulb can burn out and fool you into thinking the oven is off. **McMaster Carr part# 2779K15**

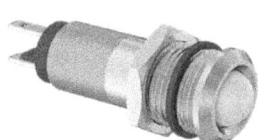

Outlet Receptacle - This is a standard 110 volt, 15 amp. wall receptacle. Make sure it will accept a three prong grounded plug. These can be found at any hardware store. **McMaster Carr part# 7195K3**

Metal Boxes - We will use one of these for the pump receptacle and two more to house the power switch, pilot light, fuse and safety timer. Use a blank cover plate drilled to accept these parts. These metal boxes come in many styles and mounting configurations, try to select one that will mount through the wood box as shown. Connect a ground wire to each box. **Mcmaster part# 71695K82**

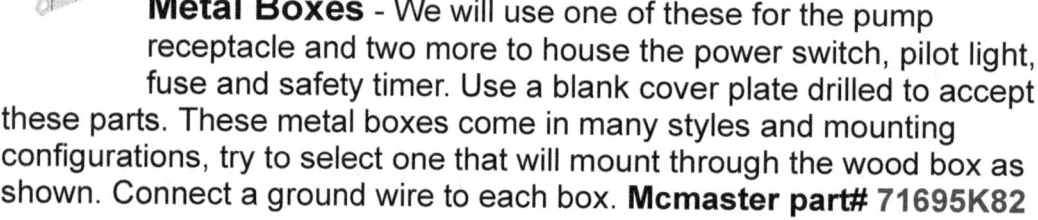

15 Amp Fuse - ¼ inch Glass fuse, find one at any hardware or auto parts store. **Mcmaster part# 7085K19**

Fuse Holder - The fuse holder shown in the photo mounts through a round hole and accepts an ordinary 1/4 glass fuse. Make sure the fuse is rated at 15 amps. The fuse holder itself will also need to have a rating of at least 15 amps, mine did, but it ran kind of warm, so I would suggest using one that has a 20 amp. rating. This fuse will protect your machine, don't leave it out! **McMaster Carr 20 Amp part# 7087K15**

Vacuum Gauge - 2 in. panel mount, 0-30 in.hg.

Qty 1, **McMaster Carr #4002K24**   **Grainger #1X501**

Caution!

-- Avoid extension cords or keep them as short as possible. Even if they have a high enough rating (15 amps.), they will cause a voltage drop and most likely result in reduced heat output from the oven.

-- The power cord will have three wires, white, black and green. There is some logic behind these colors. Green will be used for ground only, and will be connected to the oven box and outlet boxes. The white wire must always maintain a continuous unbroken connection to any load, in this case the oven and vacuum pumps. The white wire should never have a fuse or switch in its path. The black wire should first go to the fuse, then the switch and then to the loads. Follow these simple rules to minimize shock hazards.

After you have all the parts, the first step is to mount the hardware and then we can connect the wires. Drill two holes through the dividing wall into the oven compartment. These holes can be 3/8 in. dia. and will allow the oven wires to pass through. Drill them close to the side walls so you can run the wires down the sides to the front panel.

Attach the two high temp. wires that came with the oven kit to the terminal screws on your heating element. Use a nut on each side of the terminal and tighten them against each other to clamp the terminal but **not** the insulation board. Make sure the wires are pointing in the right direction so you can pass them through the holes you drilled. Be careful not to twist the heating coil by pulling on the wires. Look into the oven to make sure it is still straight.

Cut rectangular holes in the cabinet to accept the metal outlet boxes. I didn't give dimensions for these holes because your boxes may be different from mine, just cut the holes to fit closely and attach the boxes with screws. Use a blank cover plate and drill three holes to accept the power switch, pilot light and fuse holder. This plate will fit over one of the boxes and the 5 minute timer will mount in the other box on the front panel. A standard three prong receptacle will fit into the side wall of the machine to power the vacuum source. The power cord should pass through a close fitting hole in the wood box and be secured inside with a cable clamp . Round off the edges of the hole so it won't cut into the cord.

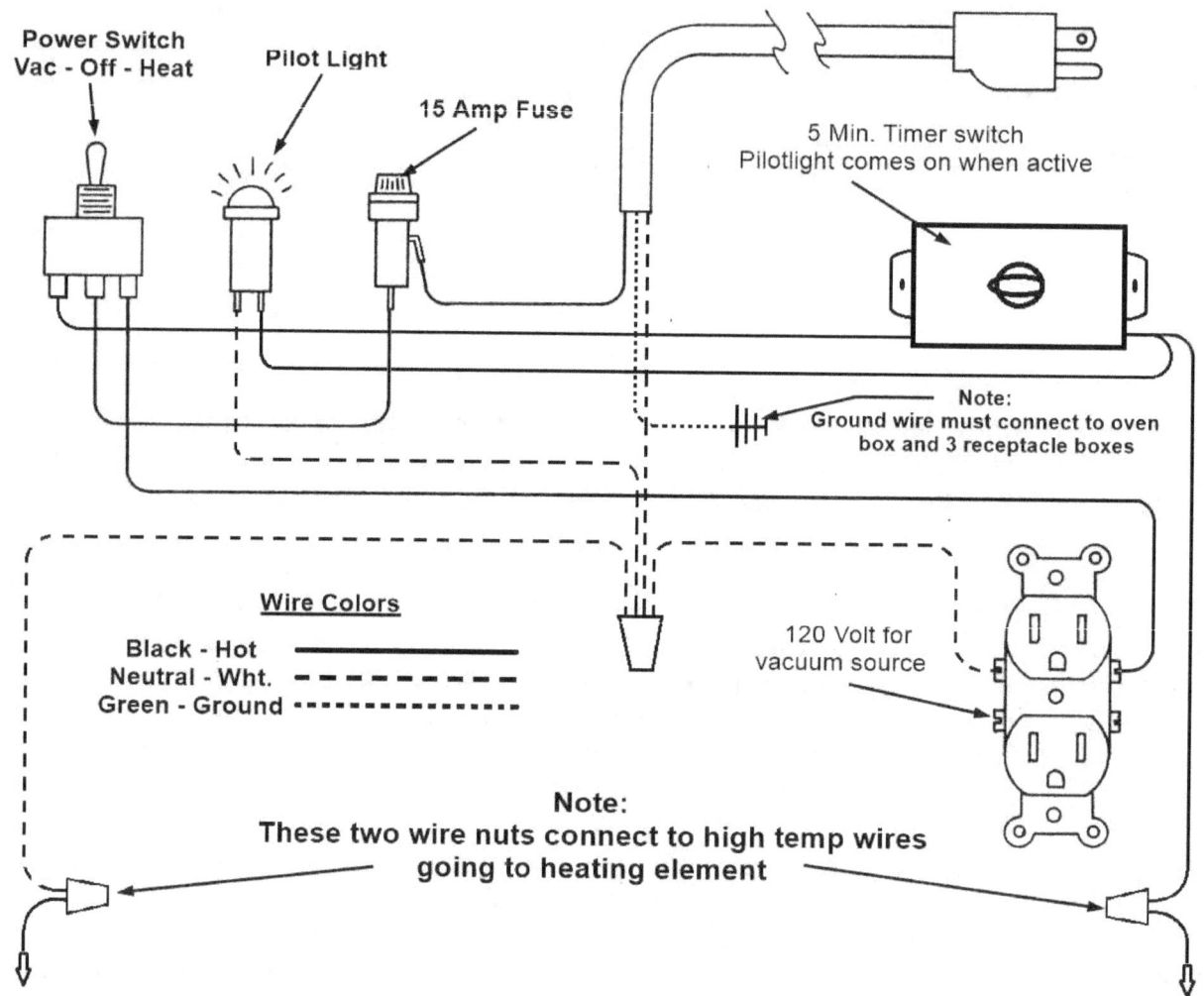

Wiring your Machine

Starting at the power cord, the green wire (ground) needs to connect to the sheet metal oven box and the three metal boxes in no particular order. There are usually extra threaded holes in the outlet boxes that you can use to ensure a good ground connection. Connect the ground wire anywhere to the oven housing. This ground wire is what will protect you if a short occurs. Use a wire nut to join all grounds together.

The white wire (neutral) from the power cord needs to go to the vacuum pump receptacle (silver screw), heating element and pilot light. The rule is that neutral wires should go to each load with no interruptions. These neutral wires can all be joined with a wire nut as shown.

The power cord black wire goes first to the fuse holder, then to the center of the power switch. Run another black wire from one end of the switch to the oven timer, then from the other side of the timer to one side of the heating element, (either side}. The oven circuit is now complete. Run one more black wire from the remaining terminal of the power switch to a brass screw on the outlet receptacle. Your vacuum pump circuit is now complete. Finish off by running a small wire from your pilot light to the hot wire coming out of the timer. It can share the same terminal with the oven wire. The pilot will only turn on when the element is energized by the timer. It will go off when it times out

Secure all wires to the wood box with small cable clamps and inspect your work. Make sure all connections are tight and then screw the cover plates onto the outlet boxes. Make sure no live connections are left exposed where you can touch them.

Test it Out

Install the bottom pegboard cover to close off the oven compartment. Turn the machine right side up and with the switch in the off position, plug it in. Make sure you have a 15 amp. fuse in the fuse holder then flip the switch to energize the oven. The pilot light and heating element will only come on after the timer knob is turned. At that point the heating element should heat up. The first time or two you power it up, the element and board will smoke as it burns in. Don't worry, this is normal and will eventually stop. Within a minute the heating element should be glowing a dull red or orange. Let it heat for a few minutes to verify that your house wiring can handle the load without tripping a breaker. The timer will time out at whatever value you set it to. I suggest 3 to 5 minutes for general use but watch for sag as an indicator of readiness.

Now test the pump circuit by plugging a vacuum pump into one of the outlets. Flip the switch to the center (off) position and the pilot light and element should turn off, then flip it into the "Vacuum" position and the pump should come on.

Congratulations, you are all wired up. If this part of the project made you the least bit nervous, I suggest you have an electrician look it over. There should only be a minimal charge for this if you take it to their place of business and remove the cover plates so everything is in plain view. They will even have a tester that can tell you the actual amperage draw. I would consider this "cheap insurance"

Warning!   Warning!   Warning!

Wow, that got your attention didn't it? I'll keep this brief but there are a few things to watch out for that may not be obvious.

-- **Don't use this oven for any other uses** - I've heard it all, people cure fiberglass, dry paint and plaster molds, warm up food, get a suntan, use it for a space heater, and a hundred other uses I don't want to know about. I am not there to stop you, but at least I warned you!

-- **Keep the machine dry** - I hate that I have to say this but I actually had a customer call and tell me his machine was left out in the rain and ask if he should plug it in to dry it. The insulation board will soak up water and stay wet for quite a while, causing a short if plugged in. Aside from normal humidity, don't operate the machine if it is wet or even damp.

-- **Clearance to Combustibles** - Use this machine only on a hard flat surface with at least 1/2 inch of unobstructed airflow underneath. Use rubber feet as shown to raise it off the table. Clear the work area of all flammable liquids and keep combustible materials such as window curtains, rags, loose papers etc... at least 12 inches from the sides and 36 inches from the top of the oven. Don't operate the machine on a carpeted or heat sensitive surface.

-- **Don't add insulation** - This design relies on the free flow of air up through and around the oven box to keep temperatures under control. The heating element itself actually exceeds 1200 degrees but a vented air gap keeps the cabinet cool. Adding insulation or omitting the vents can result in an unsafe condition.

Chapter 13 - Using your machine

Your machine should look pretty complete by now, let's go through a quick checklist to see if you forgot anything important. We will then go step by step through the forming process and finish up with a lot of helpful hints.

Assembly Checklist

-- Visually inspect the heating coil to make sure it can't possibly touch anything metal such as the brackets inside the oven box. There should be at least 1/2 in. clearance between the heating element and the oven screen even if you push on it.

-- The metal oven box, safety screen over the element and electrical outlet boxes must be grounded to the green wire on your power cord. The bottom vented cover over the oven should be in place.

-- Your main power switch should energize either the oven or the vacuum pump but not both at the same time and the center position should be off. The pilot light should always come on with the oven.

-- The machine should be plugged into a 15 amp. circuit with no other loads on it. Try to avoid the use of an extension cord or the heat output may be reduced. Use a heavy short cord if needed

-- The platen should be screwed to the top of the cabinet and the clamp frame should fit over it with equal clearance on all four sides.

-- The upper and lower clamp frames should fit together well with no appreciable gaps. Use 10 spring clips to hold full sheets in place, 8 clips for half sheets (9x12), and 6 spring clips for quarter sheets (6x9).

-- Check the vacuum system for operation and fix any leaks. If you have a vacuum gauge installed, you can test the whole system by placing electrical tape over all the platen holes. Even small leaks can make a difference.

– Find a good place for the machine with plenty of clearance to combustibles and a handy fire extinguisher. A warm room with still air is best. A cold basement or garage will cause slower heating and any kind of air movement will cause uneven heating. Using the machine outside on a carport or in an open garage, or near a fan can cause problems. You can see how sensitive it is to drafts by gently blowing on the element and watch how easily it goes dark.

Operating Sequence

Let's go through the process step by step to make a typical part. Pick an easy pattern for your first try and start with an easy to form plastic like Styrene. See the "helpful hints" section for more mold and plastic information.

1 -- Turn on the oven and let it warm up for a few minutes. The heating element will reach full heat in about one minute, but the surrounding materials need a little longer to warm up. Set your safety timer for 3 minutes. The timer is limited to 5 minutes because most plastics will take less than that. If forming ¼ inch thick, you may need to reset it once. The timer is mainly a safety backup in case you get distracted and forget to turn it off. Once you get familiar with a part, you can use the timer more for reference.

2 -- Clip the plastic sheet into the clamp frame and swing it over the oven. Reset the timer by turning it off, then on again. Set your mold on the platen

3 -- **Watch the plastic as it heats.** Do not walk away or answer the phone because some plastics heat very fast. You want to watch for the plastic to start sagging in the middle. You can poke at it with your finger to see when it gets soft. It will usually wrinkle up a bit as it expands, then go smooth again and finally start to sag. Let it sag far enough to approximate the volume of your mold, but not so far that it comes into contact with the oven screen. If left too long the plastic can burn and possibly ignite. **Remember, "Safety First" keep a fire extinguisher close by and don't take your eyes off the oven when it is turned on.**

4 -- When the plastic has the correct sag, turn on your vacuum pump and then swing the clamp frame over the forming surface. If you swing too slow the plastic can droop to the inside and form unevenly. It should take you about 1 second to complete the swing. Once flipped, put your hand through both handles and lightly squeeze them together, this will assure that the plastic makes good contact with the sealing area. Continue to hold this slight downward pressure on the clamp frame until the plastic has cooled.

5 -- If you are using a direct pump system or an automatic two stage system, then the vacuum will take over and the forming is finished in a second or two. You can verify the vacuum level by looking at the vacuum gauge. If you are using a storage tank, quickly open the ball valve as soon as the plastic has touched the platen and shut it after the plastic has firmed up.

6 -- Let the plastic tell you when it is cool enough to remove. It will usually do this by shrinking away from the seal and losing vacuum by itself. Keep the handles gently squeezed lightly together until you see this happen. You can also poke the corners with your finger to test its firmness. Release the handles to allow the vacuum to bleed off, then, switch off the vacuum system. If you turn the pump off while holding full vacuum

The pump may shudder as it comes to a stop under load. This is harmless, but make sure it doesn't walk off the table.

7 -- Remove the clips to release the plastic from the clamp frame and remove the mold from the formed part while it is still warm. You can do this by flexing or tapping on the formed part until the pattern drops out. You can also blow compressed air along the side of the mold if it's stubborn.

That's all there is to it! You can load another sheet and make parts all day long. The whole cycle time only takes a few minutes. The fun begins and now that you have this capability, you will surely think of new uses for it.

How much should I heat the plastic

The amount of sag in the plastic sheet is the best way to determine readiness to form. If you don't let it sag far enough, the plastic may not form completely. If you let it sag too far, the plastic will overstretch and form webs or wrinkles in the finished part. Using a timer or measuring the temperature are something everyone tries but watching for sag just works better. Every batch is different so times can vary . Also the machine , mold and even the room heats up so the time will always keep getting faster. An Infra-red temp gun has a hard time too because the color, opacity and gloss can affect readings so you can't really form by the numbers. Just poking the sheet with a finger and watching the sag gets the best results. Even automated machines use a photo-electric beam that gets broken so the sheet sag triggers the forming cycle. Temperature and time just aren't as reliable as sag.

Different plastics will form differently, and you will soon develop a feel for this, but it is always a good idea to keep notes so when you get good results you can repeat them later.
It's not a bad idea to sacrifice a sheet of plastic and heat it too far just to see what happens. The most common mistake is to try forming before the plastic is soft enough. As you heat the plastic It will first expand and get wavy, then it will pull tight again and finally it will start to sag. Some plastics sag slowly and some sag very fast. *Remember, watch the plastic at all times while it is heating and don't let it sag so far that it touches the oven screen.*

Faster heating - You can speed up the heating process by placing a cookie sheet on top of the clamp frame while the plastic is heating. This traps the hot air on top of the plastic sheet to reduce heat losses and cuts the heating time by about 50%. This is especially effective when heating thick sheets (over 1/8 in.) or polycarbonate. A more permanent cover can be any piece of sheet metal cut to 12 x 18 inches with a handle in the center for convenience.

Caution:

Watch very closely because the plastic will sag much sooner when it's covered. Never turn your back on the oven when the cover is in place, and don't use the cover to warm up the machine.

Avoid Heat Buildup - If you are forming a lot of parts, your mold will absorb a little more heat each time until it gets quite hot. This also holds true for the machine and even the operator. This heat build up can damage some molds made of wood or plaster. An easy solution is to place a small house fan next to the machine so it can blow across the platen. Turn on the fan after the part is formed and let it run while you remove and reload the plastic sheet. This will keep temperatures under control.

Storage -The machine can be stored laying flat (under a bed), or standing on end (in a closet), but keep the machine dry. If you store the machine in a damp location the Insulation board may absorb moisture and cause an electrical short.

Avoid Drafts - For best results, operate the machine in a warm room that is completely free from drafts. The hot air rising from the oven behaves just like invisible smoke, so its easy to imagine that even a small draft will disturb the hot air and result in one corner or side not getting enough heat. If this happens you may be able to prop up a piece of cardboard next to the machine to act as a draft shield.

Spring Clips -The smaller spring clips will work on up to 1/8 in. thick sheets of plastic. If your part involves severe stretching, the plastic may pull out of the clamp frame, simply use more clips in those areas. Extra clips are available wherever office supplies are sold and they come in three sizes, you should use the medium size to hold sheets that are over 1/8 thick.

Congratulations on following through and I hope you put your machine to good use. If you get addicted to vacuum forming as many have, visit our website. We also offer heating elements and larger machine plans for the serious hobbyist or small business

www.build-stuff.com

Chapter 14 - Plastics and Mold Materials

Where do I find plastic sheets?

There are probably a thousand small distributors of plastic across the US. Google and try to find a local source. The most common size for full sheets is 4x8 ft and less frequently 3x6 ft. If you want full pallets you should try to buy directly from a plastics extruder. That gets you much lower prices, more available sheet sizes and colors, and fresher material. Since most plastics are an "alloy" of ingredients they can even tailor the recipe or color sometimes.

More likely,.. you just want small quantities so that puts you at the mercy of a local stocking distributor because shipping is often more expensive than the material itself. Google using search words like "Plastics distributor".. "Thermoforming supplies" .."thermoplastic sheets".."Thermoformable sheets"..."Plastic sheets" etc. If you want a specific plastic, Google that by name and shop around because the prices vary a lot and often include minimum quantities and cut charges. Some places cut for free too. If a business in your area buys a lot of the type you need the inventory is likely to be fresher and cheaper. I wish I could be more helpful than that. It's very regional so some people have easy local access and some don't. If you can buy full sheets, roll them up to fit in your car, then cut them with scissors, you can save a lot of money. If your only choice is to order by mail then having them cut to size makes sense for shipping. Many suppliers now are set up for cutting and shipping

Some plastics form better than others, but you can vacuum form a wide variety of plastics, as long as they get soft when they are heated. These are called thermoplastics and include the common types listed below. The most commonly used opaque plastics are Styrene and ABS, This is due to wide availability, low cost and ease of forming. For clear plastics PET-G wins the popularity contest for its ease of forming. Acrylic is common but very brittle. Polycarbonate is the most durable but expensive and harder to form. Plastics will be discussed more in chapter 14.

Types of Plastics

ABS	Acrylic	Polyethylene
PVC	Styrene	Polypropylene
Polycarbonate	Butyrate	PET-G

"Freshness" of the plastic can sometimes be an issue too. Many plastics will absorb water from the air. So you want to find fresh stock and store it properly. If unsure which plastic or thickness to use, look for a similar product to see what they use. Here are some quick descriptions for the most common plastic types.

High impact Polystyrene, or just "Styrene". - This is the most common and available plastic for good reason. It's cheap and available in many thicknesses. White and sometimes black are stocked by most distributors. Other colors are available by special order. This plastic forms easier than anything else with great definition, has a long shelf life with no moisture absorption issues. It glues and paints easily and should be your first choice to test a new mold or use for hobby or craft uses.

PET-G, This is a clear plastic much like soda bottles or blister packaging. It's the friendliest clear plastic with great durability, ease of forming and no moisture issues. It's also easy to find in many thicknesses. It can also be tinted with dyes. This and Styrene are the favorites of modelers and hobbyists.

ABS - This Is a blend that offers excellent durability, formability and is plentiful and fairly cheap. It's mostly available in black, sometimes white and rarely in other colors. You can get it with a smooth texture or a popular "hair cell" texture similar to leather. It's a bit more durable than Styrene and has better weatherability when not painted. The texture looks nice too. The negatives are that it's not as consistent because it has a high recycled content and each vendor blends it differently. When you get a good batch it's really formable. Other times it's less formable and has moisture problems. You have to accept this variability along with a higher scrap rate. Store ABS in a low humidity air conditioned room. It is possible to pre-dry sheets with moderate moisture problems. "Google Pre-Drying ABS" and the manufacturers name if you know it, They all publish forming and fabricating guides for their products.

PVC - This is another plastic that can be great to work with, or not so great. It's a little less easy to find but it forms, glues,and paints well. Being another blended plastic its formability can vary but moisture isn't a problem, Available in gray or clear. The clear isn't as user friendly as PET-G so try that first for most things,

Polycarbonate - A common brand name is "Lexan". This plastic stands out in several ways. It's by far more durable than all others. It's very easy to find but it's a lot more expensive, it heats at higher temperature and chills more quickly. These things make it tricker to work with. On top of that it has bad issues with moisture. If you can get it fresh, store it properly and use it fast, its not so bad. If you need crisp definition, you may need to pre-heat your mold because the plastic chills or cools much quicker so you need a good oven and a lot of vacuum very fast to form it well. Get it all right and it's worth it If you need the toughness.

Acrylic - Although it's super common in clear and available in many transparent tints, it's a very brittle plastic. Because it's so hard it can be sanded and polished so it's good for windows, tail light lenses and things that need to be looked through. It forms fairly well but cracks easily when being trimmed.

There are many other plastics but the above choices are easiest to find and use. You can find thermoforming guidelines from the manufacturers for most of these materials online. One final comment on drying ABS and Polycarbonate. Again, the manufacturers publish more details but the concept is to heat the plastic at a lower temp to evaporate the moisture before forming. You can make a low temp drying box but the temperature and time are dependent on how bad the problem is. Some dry out quickly, others can take overnight. Some may never be usable if it's gone too far. You can never know how old the plastic is because distributors just stack more on top when they get it and the bottom sheets may be years old. Age affects all plastics but at different rates. The plasticizers can outgas over time affecting the properties too. Generally the plastic gets more brittle and less pliable as it ages

2

.Mold Types and Materials

Part of the reason vacuum forming is appealing is because you can form over many things. A mold just has to be firm enough not to compress or distort and it only has to endure contact with the hot plastic for a short time. Wood, Plaster, Epoxy, Plastic, Metal, Ceramic, even cold modeling clay can be used for one part. That's the best part about this process, a mold for injection molding might cost $20.000 and a mold for vacuum forming might cost 20 cents. Most people use plain old wood to develop the shape and then cast an epoxy replica for extended use. A solid cast resin mold is affordable and can still produce thousands of parts. Lets discuss the range of things we can use as molds.

3D printing - This is the trendy thing right now and it is an ideal companion
To vacuum forming so I'll devote the final chapter to this one. All the tips and rules given here will also apply.

CNC machined molds - Similar to 3D printing, it allows you to create and design on your PC then create an accurate shape. Aluminum makes a great mold but may need to be pre heated or even cooled for long runs. There are some great machinable tooling boards with properties like dense wood or foam and some are high temp as well. If you have access to CNC machining its a great way to make permanent tooling. Of course the machinery and tooling boards can be expensive.

Found items - These are the easiest,.. Just grab something with the right shape and form over it. The introduction covered some limitations on shapes and proportions but many hard sturdy objects can work. With thin plastics the heat isn't too bad and it cools quickly but you don't want to form over wax or some foams that melt easily. You can also crush objects if you are not careful.

Simple Wood Molds - I can't imagine an easier way to make custom molds. Simply carve or sand a chunk of soft or hard wood to the shape you want and form over it. No paint, no mold release, no big deal, just try to avoid woods with an open grain such as oak or ash. You can get dozens of parts off of bare wood molds. You can even glue little chunks together to make big molds, but don't use a hard glue because it will sand unevenly. I use auto body filler (Bondo) as a glue because it sands easily and sets up fast. I use white styrene to make model airplane parts over wood molds and very little wood grain shows through the plastic. You can always wet sand the formed part before painting to remove imperfections.

If you want the wood molds to last longer, you can brush on several coats of ordinary fiberglass (polyester) resin, sanding between coats. This will put a hard shell on the wood and also help hide the grain. Epoxy resins do not fill as well and sometimes blister from the heat and stick to the plastic.

Plaster - This is what comes to mind first for most people and it does work OK. There are many types of plasters with different hardnesses. Plaster of Paris is cheap and fairly soft with a smooth texture. Hydrocal is a harder plaster. The attraction is that they are readily available, cheap and easy to use. The downsides are that they need to be very dry to develop good strength and even then they can hold enough moisture to sweat and corrode your platen or suck moisture and dust into your vacuum pump. Plaster is also prone to chipping and cracking. It's also very heavy for large molds. It works best for short runs or one time use because of its low cost.

Cast Resin Molds - A plaster or wood mold works great, but will deteriorate when used too much. If you want to go into mass production then consider casting permanent molds from a two part resin. You can start with a simple wood mold to prove a concept and refine the shape. Once its final, then cast a resin copy for production use and store the wood master as a backup. You can use a silicone mold off the wood, or just vacuum form a thin styrene part to use as a casting mold. Choosing the right resin can be a complicated subject and there are a million products available in both Urethane and Epoxy formulations. The two companies below offer excellent products for making vacuum forming molds.

Polytek Development Co.
55 Hilton St.
Easton, PA. 18042   (610) 559-8620
www.Polytek.com

Ask for BC-8009 Slo-Kast gray urethane
This is my personal favorite. It's by far the easiest to use and the cheapest. This is a two part urethane that mixes 50/50 with one part black and one part white so it turns a

uniform gray color when mixed. The viscosity is thin so it releases air bubbles and pours easily. The cure time is 3-4 hours and the pot life is a reasonable 14-18 minutes. It mixes so easily and doesn't need de-gassing. It can be drilled and sanded and is a nice stable material with low shrinkage.

Smooth-On Inc.
5600 Lower Macungie Rd.
Macungie, PA. 18062   (800)762-0744
www.smooth-on.com

Smooth-On makes a urethane called Task 18 that is ideal for molds. They also offer a vast selection of molding and casting products and are known for their great customer support and seminars along with an informative website.

Of course Google can lead you to many more sources for quality casting resins.

Chapter 15 - Making your molds

You will need to have realistic expectations about which shapes are formable and which shapes can cause webbing thinning or failure to come off the mold. Please refer to page 122 for a helpful diagram showing the size and shape limits for this machine. With that in mind, the following helpful tips apply to all molds.

Some simple rules to follow.

— Your mold can not have undercuts. These are areas where the plastic can wrap around or under the pattern in such a way that you can't remove the formed part.

— The plastic will shrink as it cools. This shrinkage is in the range of .005 to .008 in. per inch. This can be as much as 1/8 inch. over a 16 inch length. If you can't tolerate this then you must make the pattern oversize.

— You should always provide a slight angle on vertical sides of your pattern. This is called "Draft" angle and should be at least 3 to 5 degrees per side. The reason for this is the shrinkage mentioned above. If you have tapered sides then it will come off more easily.

— The wall thickness will not be uniform on a finished part. The plastic will be the thickest where it first touches the pattern and the last part to touch will be the thinnest. The amount of thinning depends entirely on how much the plastic has to stretch.

— There is a general rule of thumb that the pattern should not be taller than it is wide. This can result in webbing (wrinkles) in the plastic usually coming from a tall corner, as well as excessive thinning. This depends a lot on the shape of the pattern with rounded shapes being better than square ones. The pattern should also be placed at least as far from the edge of the platen as it is tall. If it's too close then the plastic can get thin or not seal to the platen. Be realistic with the size and shape of your mold

Mold making tips

Making Multiple Molds - Let's say you made a great little hand carved mold for slot car bodies and you want to form 10 parts at a time. You obviously don't want to carve nine more molds. Here's a trick for reproducing the original one. Vacuum form a high definition part from your original pattern, by this I mean use a thin sheet of plastic that forms easily such as styrene and use maximum vacuum so it forms crisp detail. Now use this part as a mold to cast as many duplicates as you need from plaster or a two part resin. Save the original wood pattern as a master and use the cast ones for production.

Female Molds - When you vacuum form over a male mold, the plastic tends to obscure some of the detail. For example, you can look inside the part and see the wood grain or texture left from the mold, but the outside is smoother. This works to our advantage sometimes if the mold is less than perfect. Other times, you want the maximum detail possible on the outside of the part. You can achieve this by forming into a female mold or cavity, Your equipment doesn't care which type of mold you use.

Female molds can be made by first creating a male pattern and then casting plaster or resin over it to form a block of material with a reverse impression or cavity in it. You will have to drill some small holes in this cavity so the air can get sucked out, but don't get carried away, most people drill way too many. Wherever you drill a hole (generally 1/32 or smaller), it will leave a small raised bump on the outside of the part that can easily be sanded off. This technique is rarely used but can produce impressive detail.

No Undercuts - This one should be common sense but it's amazing how many people don't get it. If you vacuum form over a tennis ball, the ball will never come out because the plastic wrapped too far around it. This is a severe undercut. Examine the sides of your mold for any depressions or places where the plastic can get sucked into. You may have to fill these areas or your mold will never release.

Make friendly molds - The plastic sheet will shrink up to 1/16 in. per foot as it cools. You can imagine that this will cause it to grip your mold tightly, especially if it has straight sides. A friendly mold would have sides that taper at least 5 degrees and nice round corners for the plastic to slide over as it is being stretched. I know you can't always follow these rules, but the mold will release easier if you do.

Webbing -This is where the plastic folds together and forms creases. It can be caused by over stretching the plastic (too much sag), but the shape of your mold may also be a factor. The general rule of thumb is that you can form parts that are no higher than they are wide, but shapes with vertical sides and square corners are more prone to webbing than ones with tapered sides and rounded corners. Webs can also occur when multiple molds are placed too close together. Here are some general tips to prevent webbing:

-- Try less sag before forming, this creates more tension in the plastic sheet.
-- Thicker plastics will have less tendency to web than thinner sheets.
-- Place scrap blocks of wood on either side of the area that is webbing, by varying the size and position of the blocks you can usually cause the web to flatten out or sometimes trick it into webbing off your scraps instead of your mold.
-- Modify the mold into a friendlier shape for vacuum forming.

Avoid Hollow molds - This topic doesn't get enough attention but it affects a lot of things. The air space inside a hollow mold adds extra volume that must be evacuated,

this can slow down your vacuum pump so much that the plastic will cool before you completely form the part. Hollow molds may also collapse or damage the machine. This is more true if your vacuum system is strong. For example, say you have a heavy steel pan and you want to vacuum form over it to make some storage bins. If the pan measures 10 x 16 x 2 , that's 320 cubic inches of air to evacuate which will take way too long. But that's not the scary part, The inside of the pan has 232 square inches of surface area, so if you can pull a 27 in.hg. vacuum (equal to 13 psi), this will result in over 3000 lbs. of force trying to pull the pan down to the machine. You **Will** crush the pan and maybe damage your platen.

Now this is an extreme example, and you can get away with a hollow mold if it is small and sturdy, but it should have ribs or bracing so it doesn't damage the platen.You can also back fill the mold with something like plaster or concrete.

No Mold Release - For the longest time I was using various slippery substances on my molds just because I thought the plastic might stick, but it usually doesn't. Now I rarely use anything at all, try forming without mold release first. Then use it sparingly if needed. Silicone spray or paste wax work well but powders or oils can get sucked into your pump and eventually cause damage

There is also a strong urge to want to "seal' your wood or plaster mold with paint or varnish. These coatings can actually stick to the plastic more when hot, if you want to fill in the wood grain and make your molds last longer, coat them with a couple coats of ordinary fiberglass (polyester) resin. I find that epoxy resins, and enamels can tend to get soft and sticky when hot. If you don't have a sticky coating but are having a problem removing your mold from the part, it's most likely a problem with the shape, not the plastic sticking. The plastic shrinks considerably when it cools and will grip your mold tightly. You can blow it off with compressed air, or remove it while it is still hot. As a last resort, try making a cut in the plastic where it can be repaired later. If you are still having problems, try a silicone spray, furniture polish, paste wax, vaseline or just about any kind of oil or wax.Just don't use so much it gets sucked into the pump.

Shim Up Your Mold - If you place a large mold on your machine, be careful not to restrict the air flow too much by covering all the holes. You can space the mold up about 1/16 in. simply by placing coins under it no farther than 2 inches apart, or better yet a piece of 1/2 in mesh screen.

Don't over polish Your Mold - If you are making clear parts, resist the temptation to polish your mold to perfection. This will almost always cause the hot plastic to seal off and trap large air pockets against the surface. It's better to use 400 to 600 grit sandpaper or steel wool to scuff it up slightly and provide tiny air channels. This lets the air pockets bleed down yet still results in a clear part.

Clear Parts - Use this trick to make high quality clear parts without spending a lot of time making perfect molds. Great for airplane canopies and other windows. Vacuum form a sheet of thin styrene .020 to .060 over your bare wood or plaster pattern. This should hide all of the surface texture. (thicker plastic hides more) You can even wet sand the outside of the styrene with 400 or 600 grit paper to remove any imperfections. trim out this part and place it back over the wood to act as a smooth shell, then form your clear plastic over it with no mold release. If you are using clear Pet-G you can re-use the styrene shell several times. If you are using Polycarbonate, you will have to make a new shell for each clear part. I have made many airplane canopies this way. A simple wood mold covered with thin styrene and wet sanded. This makes a surprisingly clear part with minimal effort.

An example of best practices

This example is a fairly deep draw but is also a friendly shape. Notice that it's as tall as it is wide which starts to push the limits however the friendly rounded shape with no undercuts makes this an easy part to form. This is a good example of best practices and how they can be the difference between failure and success. This was made on a 6 x 9 inch platen.

Look at any vacuum formed parts you can find and you will see that they follow these simple rules. Despite these limitations there are millions of uses for vacuum forming. Perhaps the greatest advantage to this versatile process is the low tooling costs for the pattern. To make an 8 x 12 inch part with injection molding you would have to spend around $20.000 for a tool and wait 8 to12 weeks. With vacuum forming you can make a simple wood pattern for a couple of dollars, Or a permanent epoxy pattern for around $25.00, all in one afternoon.

What are the Practical Limit's of vacuum forming

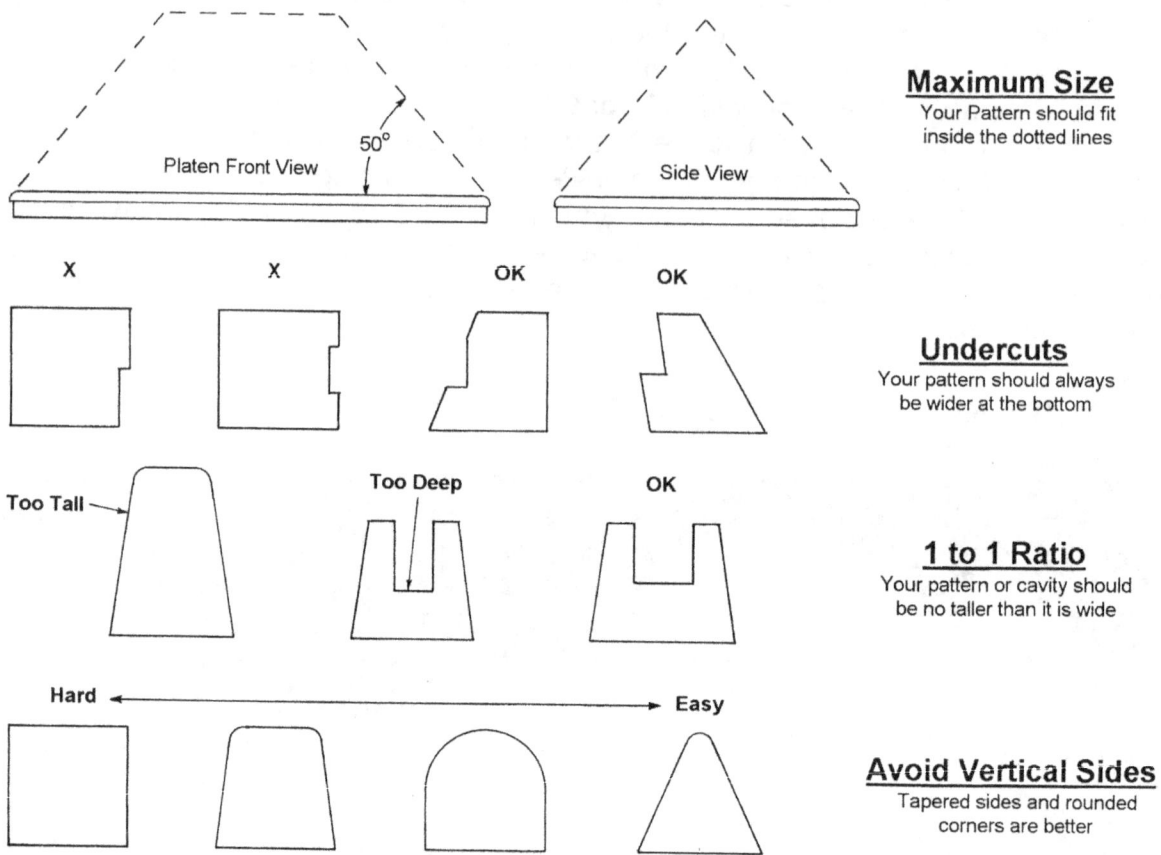

Maximum Size
Your Pattern should fit inside the dotted lines

Undercuts
Your pattern should always be wider at the bottom

1 to 1 Ratio
Your pattern or cavity should be no taller than it is wide

Avoid Vertical Sides
Tapered sides and rounded corners are better

The 1 to 1 ratio of width to height should be considered maximum. When possible stay below these limit's and your success rate will be higher. There are advanced techniques to enable deeper draw ratios, but they are beyond the scope of these plans. There are some inherent limitations to the vacuum forming process no matter how much you spend for the machine and some shapes are just not suitable for vacuum forming. The drawings above illustrate some good and bad shapes and show the maximum pattern size you can use.

Anything you can do to taper the sides and round the corners will make it easier to form. Square corners and vertical sides are the hardest to form. The plastic shrinks when it cools and will grip your pattern tightly. Look at any vacuum formed parts you can find and you will see that they follow these simple rules. Despite these limitations there are millions of uses for vacuum forming. Perhaps the greatest advantage to this versatile process is the low tooling costs for the pattern.

Chapter 16 - 3D Printed molds

I like to call vacuum forming the original rapid prototype method. Now we have 3D printing as a way to rapidly make a mold for vacuum forming.The two machines when used together are what I like to call "Desktop Manufacturing".3D printing is a great way to make molds. There are still ways to do it wrong and you still have to obey all the same limitations regarding shapes and undercuts as described in the introduction. It's not a magical solution but it is one more good way to make a mold. The same can be said of CNC machining. If you have that equipment and the skill to use it. Some people have more of an aptitude for "hands on" carving, sculpting or woodworking. Younger readers tend to lean more towards visualizing or creating on their computer. Vacuum forming machines don't really care as long as the mold is a friendly shape, is sturdy enough and can withstand some heat for a short time. The hot plastic and vacuum will do its best to replicate any shape and texture it touches.

3D printing is a broad term covering many methods and materials. Surface finish or resolution is the first thing you notice so finer resolutions or more detail is better. CNC machining is the same because it leaves fine ridges and cutter marks. Its a good idea to print, then vacuum form a test part for evaluation and then decide whether to put more work into sanding or filling imperfections. All of the 3D printing choices can make at least a few parts but some can last much longer. There is a world of info on 3D printing and its many variations so I won't go too deep. Lets just address the major methods and how they relate to vacuum forming.

Filaments - Liquid resin - Powdered resins

FDM - or "fused deposition modeling" Uses a plastic filament which is extruded in layers to build up a part. This is the most popular method for hobbyists. Affordable supplies and machines come in useful build sizes. This process suffers from fairly coarse resolution or layering but is improving with finer nozzles. You can address this with sanding or filling but be warned this can take some time. A brief description of filament types is as follows...

– PLA filaments are the easiest but are fairly low temp and low strength material. You can get at least a few good vacuum form pulls if you use a thick shell and dense infill.

– PET-G is another easy but low temp material with better durability. PLA and PET-G are the only two you can use with a cold forming bed and open chamber printer.

– ABS has higher temp resistance but prefers a closed chamber with heated bed to control shrinkage and warping. It can be solvent treated to some degree to melt off some of the ridges and smooth it up a bit

– Nylon and carbon filled nylon have great strength and temp resistance but require a quality machine with a high temp print head, heated bed and chamber.It also absorbs moisture and may need to be dried first.

– SLA or " Stereolithography" printers use a UV cured liquid resin that provides a much better surface finish and crisp detail than FDM. SLA is a bit slower and usually has a smaller build area than FDM for a given price point. The materials are more expensive, smellier and have a shorter shelf life than filaments. You need to wash and final cure the finished parts. The resin choices are more extensive, some are brittle and others tough or flexible. There are some very high temp options too and molds can be printed solid or hollow. They can't do honeycomb infills but can use thick walls and ribs for support. Overall a better end result is possible compared to FDM.

– SLS or "selective laser sintering" uses a powdered resin cured by lasers. This process requires the part be scrubbed to remove uncured powder and given a final cure. The surface is very uniform with no layer lines. It has excellent detail and clean corners but has a fine sandstone like texture. SLS machines are very expensive. They are faster and the parts are tough and usable so its a good choice for low volume production or working prototypes.. Plastic choices are limited but tend to be tougher plastics like nylon. The textured surface finish is fine for vacuum forming if you are OK with the look.

Helpful Tips

When filament printing you should increase the shell thickness or layers to 3-5mm. Also increase infill density to 25-50% A thinner shell can soften and allow the honeycomb infill pattern to show through at the surface.

Remember hollow molds are never ideal. If the molds are small and sturdy you may get away with it. Large volumes inside a mold can slow the evacuation time too much. Always use a dense infill or thick shell with supporting ribs on hollow molds. If you still have problems you can backfill with a liquid casting resin to make it solid. You can bulk up the backfill resin with a dry filler such as sand to reduce resin costs.

Your vacuum formed part will always shrink aggressively no matter what mold is used. With coarse surface finishes from print layering or powders, you may want to increase draft angles to help the parts release.

If dimensional accuracy is important, you will have to build in shrink allowances. The 3D printed mold and vac formed parts will all shrink as they cool.

A good strategy may be to do your rapid development with simple FDM prints, then if a higher quality mold is needed you can send the file to a print service to get an SLA or SLS mold made, or make a solid resin cast copy from the mold or one of the formed parts.

We also sell plans for a larger machine that is perfect for serious modelers or small businesses

Can be built in three sizes 2x2 ft. 2X3 ft. 2x4 ft.

Finally an affordable solution for the thousands of hobbyists and small businesses that need vacuum formed parts. Now the alternative to those high priced machines is to build it yourself. Designed to use "off the shelf" components wherever possible and simple construction with no machined parts. Our infra-red heating element kits are optimized to guarantee top performance and speed up construction.

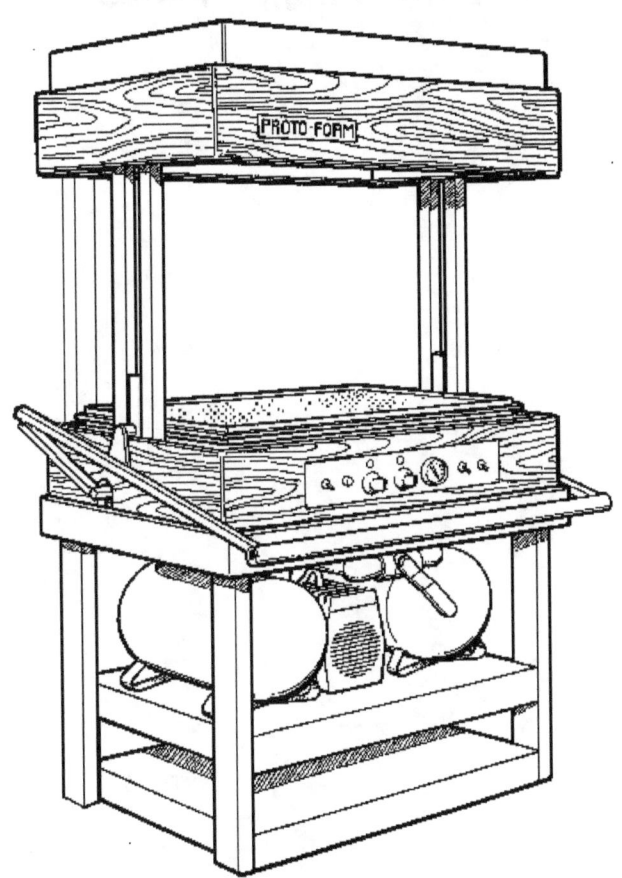

Vacuum Forming Machines

Build it yourself and save !!

Please visit our website for other downloadable plans and heating element kits

www.build-Stuff.com

hobbyvac@yahoo.com

(248) 391-2974

Workshop Publishing
2909 Crozier Rd.
Lupton, Michigan 48635

Notes:

Notes:

www.ingramcontent.com/pod-product-compliance
Lightning Source LLC
Chambersburg PA
CBHW080844120626
46553CB00009B/2557